ARGUMENTS ON EVOLUTION

ARGUMENTS
ON EVOLUTION

A PALEONTOLOGIST'S PERSPECTIVE

ANTONI HOFFMAN
Institute of Paleobiology
Polish Academy of Sciences
Warsaw

and

Lamont-Doherty Geological Observatory
Columbia University

New York   Oxford
OXFORD UNIVERSITY PRESS
1989

Oxford University Press

Oxford   New York   Toronto
Delhi   Bombay   Calcutta   Madras   Karachi
Petaling Jaya   Singapore   Hong Kong   Tokyo
Nairobi   Dar es Salaam   Cape Town
Melbourne   Auckland

and associated companies in
Berlin   Ibadan

Library of Congress Cataloging-in-Publication Data
Hoffman, Antoni.
   Arguments on evolution.
   Bibliography: p.
   Includes index.
   1. Paleontology.   2. Evolution.   I. Title.
QE721.H64   1989   560   88-9973
ISBN 0-19-504443-6

9 8 7 6 5 4 3 2 1

Printed in the United States of America
on acid-free paper

1-24-91

Preface

This is a book of unabashed criticism.

Even a perfunctory reading of recent academic books and journals devoted to evolutionary biology shows how large is the number of new hypotheses and theories proposed in the last dozen years or so as challenges to the established orthodoxy—the neodarwinian paradigm of evolution. It also shows that to challenge this paradigm has become a way to gain scientific fame and fortune, to achieve an intellectual authority. Clearly, the new is fashionable in evolutionary biology, and rightly so. New ideas are necessary in science. They in fact constitute what science is all about. For science is the incessant questioning of old explanations for natural phenomena and the equally incessant searching for more adequate explanations. Challenges to old paradigms are therefore essential to any science. This is particularly true in a science like evolutionary biology, where the established paradigm is indeed relatively old and apparently rather amorphous.

The criticism that pervades this book, however, is not directed against neodarwinism. I am a paleontologist, interested primarily in questions of biological evolution. Today, the greatest authorities in my field of scientific enquiry, especially in America, are associated with very spectacular attempts to overturn, or at least to substantially revise, neodarwinism as the way of looking at evolution. They have been so successful that the neodarwinian paradigm seems to be regarded by many of my fellow paleobiologists as deadwood that must be cut before science can progress any further. Yet I have grown up in a tradition of questioning all authority. Perhaps this is a byproduct of the school that, as a rule, expected us, students, to believe without any second thoughts whatever the authority—the teacher, the textbook, you name it—gave us to believe, while it was often quite obvious that the authority was wrong. Skepticism has thus become my nature. The accidents of my personal life, moreover, have taken me from the University of Warsaw in Poland, to the University of Tübingen in West Germany, the University of Wisconsin at Madison, and Columbia University's Lamont-Doherty Geological Observatory in the United States, and then back to Tübingen and to the Polish Academy of Sciences in Warsaw. At each of these institutions, as well as during many shorter trips I had an occasion to make, I met colleagues who were very critical of one or another tenet of the neodarwinian paradigm or of one or another challenge to it. They very generously shared their time and ideas with me and they often convinced me of the justice of their views, but since their arguments very often clashed with one another they have ultimately sharpened my skepticism even further. This book is an expression of this fundamentally skeptical attitude in a situation where the authorities in my discipline of science seem to no longer accept neodarwinism as the paradigm of evolutionary biology. It is aimed to review—from this explicitly personal perspective—the arguments on evolution that have raged over more than a decade in paleobiology.

There are plenty of debates in modern evolutionary biology. I am a paleontologist, however, and I have to focus my attention on the issues that organize the research in my field. This is the area of my competence as a scientist—and also as a skeptic. Perhaps this book would not be as critical of new theoretical developments, challenges to neodarwinism, if I were an ecologist or systematist. I tend to think it would not be if I were a geneticist or developmental biologist. But I am a paleontologist, and it just happens that I am convinced by next to nothing in what has thus far been presented in this field as arguments against the neodarwinian paradigm. The message of this book, therefore, is that insofar as paleobiology is concerned, neodarwinism, as I understand it, is at the moment well and healthy, instead of being fundamentally flawed or at least evidently incomplete.

It is the essence of science, however, that such assessments may turn out to be wrong even before they are printed. This apparent afterthought, which might be taken as a mere safeguard against betting without any reservations on a wrong intellectual horse, is actually the other important message I would like this book to convey. Except for pure logic, nothing is certain forever in science, and especially in a historical science like paleobiology. This is so not only because new discoveries, new techniques, new perspectives on old data stimulate new hypotheses and theories to explain natural phenomena, but also because new methodological and philosophical options suggest acceptance of other—no matter whether old or new—explanations.

In the last decade or so evolutionary biology has not only been a stage for hot debates about the laws of evolutionary processes and the course these processes did actually take on the Earth; the fundamental tenet accepted by all participants in these discussions—namely, the belief that species evolve and give rise to other species—has also been seriously questioned. The opening chapter of this book is therefore intended to explain the status of biological evolution as I see it. After this Prologue, I describe the background of the current debates on evolution in paleobiology. In Chapter 2, I survey recent developments in evolutionary biology and place the paleobiological challenges to neodarwinism in the context of questions and doubts raised in other biological disciplines. Chapter 3 then presents the neodarwinian paradigm of evolution, or rather what I think should be understood under this heading. There are nearly as many interpretations of neodarwinism as there are evolutionary biologists, and I do not pretend to offer here either a consensus, or the most appropriate view. I believe, however, that my interpretation is both internally coherent and sufficiently broad, though not all-encompassing, to be consistent with the opinions of the majority of those who call themselves neodarwinians. Chapter 4 outlines the properties of the fossil record that determine the quality of paleontological data and hence must always be kept in mind while evaluating any paleobiological inference about evolution. This information on empirical limitations inherent in all paleontological research is then implicitly assumed in the chapters devoted to detailed discussion of the paleobiological arguments on evolution. All research, however, is also inevitably conducted within a methodological framework. Paleontology differs in this respect from the majority of other biological disciplines in that it is fundamentally historical in nature. It deals with historical data and aims to reconstruct the historical process of evolution. A brief sketch of what I regard as the methodological context of evolutionary paleontology as a historical science is presented in Chapter 5.

The majority of the book is devoted to the issue of macroevolution, which I discuss in two separate parts—under the headings of

Macroevolution and Megaevolution. Macroevolution refers to all supraspecific patterns of evolution; megaevolution encompasses a subset of these phenomena, namely those on the grandest scale. It is one of the main questions of modern evolutionary paleontology whether or not macro- and megaevolutionary phenomena call for specifically macro- and megaevolutionary explanations, respectively. I introduce the problem of macroevolution, in its modern version, in Chapter 6. Then, I go on to discuss in detail the concept of punctuated equilibrium in Chapter 7. I argue that the available formulations of this explicitly macroevolutionary concept allow for a variety of interpretations, but under none of them do I perceive any reason to postulate inadequacy of the neodarwinian paradigm, as it has been repeatedly done by the advocates of punctuated equilibrium. Chapter 8 briefly presents the ongoing discussion on the logical status of the biological species. This philosophical issue figures quite prominently in the debates on macroevolution, though I think it is a question without much relevance for biology. In Chapter 9, the macroevolutionary theory of species selection is outlined and evaluated. I argue that although species selection, when properly defined, may potentially occur in nature, there is at present no good example of its actual action. It is therefore premature to accept it as a real mechanism of evolution.

The distinction between macro- and megaevolution is spelled out in full in Chapter 10. At present, mass extinctions constitute the most spectacular field of research on megaevolution. I discuss these phenomena in Chapter 11 and conclude that they are perfectly compatible with neodarwinism. The focus on mass extinctions certainly results in fascinating insights and information about the course of evolution, but not in a need for revision of the neodarwinian paradigm. Chapter 12 is devoted to the debate on the pattern of changes in taxonomic diversity of the biosphere during the Phanerozoic. There are several rival models aiming to explain this megaevolutionary pattern, and my discussion intends to point out the reasons I have for supporting a model compatible with the neodarwinian perspective on evolution. I hope to show that the stance I take in this debate on biotic diversification stems not as much from irrepressible partisanship for neodarwinism as from my explicit methodological options and from my sober assessment of the quality of paleontological data.

Thus, I conclude in the Epilogue that no order has thus far been discovered in the historical pattern of evolution, and the absence of any order suggests the absence of any laws of the historical process.

This conclusion implies, in turn, that all new theoretical developments in evolutionary paleontology in the last dozen years or so have failed as challenges to neodarwinism, though they have tremendously advanced our understanding of the course of biological evolution—the history of life on the Earth—and of the potential and limitations of the methods we must employ to improve this understanding even further. There is in my view no better interpretation of macro- and megaevolution available on the marketplace of scientific ideas than the following corollary of the neodarwinian paradigm: all macro- and megaevolution is an outcome of the interaction between microevolutionary processes operating within myriads of individual populations and species, on the one hand, and the global environment evolving on various continents and in various seas and oceans over millions of years, on the other. This corollary may be inelegant as a solution to a scientific problem, because it boils down to denying the need for explanation of large-scale historical biological patterns by asserting that they must simply reflect a summation of smaller-scale phenomena that find adequate explanations within the neodarwinian paradigm. I believe, however, that it should be provisionally accepted—that is, until new evidence suggests us to reject it as inadequate, or new macro- or megaevolutionary ideas appear that better explain the phenomena, or new methodological criteria are adopted that advise us to make other choices between competing hypotheses or theories. This is, then, the positive side of my critical argument: all evolution results from interaction between living systems and environment, and none of these components must be neglected in our attempts to explain historical biological patterns.

This book would never be written had not some of my older colleagues enchanted me, several years ago, with their revolt against neodarwinism. I have then rebelled against them. The ideas advocated by Niles Eldredge, Stephen Jay Gould, David Raup, Jack Sepkoski, Steven Stanley, and Elisabeth Vrba are the prime targets of my criticism in this book. I wish to gratefully acknowledge here my indebtedness, for I developed my own perspective on evolution in a continual debate. I have greatly benefited from many, often very long discussions with Stefan Amsterdamski, Peter Bretsky, Jerzy Dzik, Karl Flessa, Anthony Hallam, Max Hecht, Józef Kaźmierczak, Jennifer Kitchell, Adam Łomnicki, John Maynard Smith, Wolf Reif, Marcin Ryszkiewicz, the late Thomas Schopf, Adolf Seilacher, and Nils Stenseth. None of these colleages is likely to agree with all—and sometimes not even with the majority—of my arguments, but all of them have considerably contributed to their development. I owe special thanks to three friends. The first draft of this book was read by Jan Kozłowski, Krzysztof Małkowski, and Jerzy Trammer, whose critical comments have made me improve it substantially. Needless to say, I alone am responsible for the failings that remain.

Warszawa, Poland   A. H.
Summer 1987

Contents

ARGUMENTS ON EVOLUTION

PROLOGUE

1 Why Accept Evolution?

These are exciting times for us, evolutionists. Reports of new concepts, models, and hypotheses, of new discoveries and reinterpretations of old observations appear all the time. Consequently, evolutionary biology is in a healthy state of continual flux. Sometimes, the new empirical and theoretical evidence may help to settle some old controversies; more often than not the new material tends to reinvigorate old debates and to provoke new ones. These arguments on evolution constitute as much of the thrill of our research as does the filling of particularly substantial gaps in our empirical knowledge.

We argue about evolution in all its aspects. What is the pattern of life—at all levels of biological organization, from molecular to supraspecific—and how and why did it originate? What are the mechanisms of evolution that operate at all these levels? What is the relative significance of particular evolutionary forces and the nature of their interplay? How are they all constrained by the evolutionary history of all biological entities and by the boundary conditions of the evolutionary process? How does evolution relate to biogeography, ecology, embryology, and biochemistry; alternatively, what is its importance for these realms of the biological inquiry? And, perhaps most important, how can we best approach all these problems? To be sure, we believe we know at least partial answers to these and many other questions, but even so there is more than enough room for argument.

The one thing about evolution that we do not need to argue about is validity of the concept of biological evolution itself. We all agree that evolution does occur in nature, that species arise from other species and evolve through time, and that the biosphere as we see it today is the product of a long history of biological evolution. And we feel that this consensus reflects upon something more than merely the commonality of our vital interests as evolutionists. Certainly, we all make our living, and we hope to make also our fame and fortune, by accepting this consensus and by working within the conceptual framework it imposes upon our worldview. But we also are fully convinced that even after a soul-searching session, after a most honest attempt to strip ourselves of all selfish motivations, we still would reach precisely the same—and therefore a genuinely objective—consensus. And we strongly believe that everyone else should also agree with us on this single point.

Yet this fundamental point about biological evolution is also being contested in these times. The concept of species evolving from other species—which is often misleadingly called the fact of evolution, in order to distinguish it clearly from the theory of evolution—is rejected by creationists of diverse convictions, particularly in America but also in Europe and presumably all over the world. Creationists argue that this concept is merely an unsubstantiated presumption and hence, that the logical status of the science of evolution does not differ from the account of Creation as given by the Bible; for, in this view, they both are based on nothing but a leap of faith.

We as evolutionists might choose to simply ignore the creationist challenge because it is founded on a very serious misunderstanding, if not a willful distortion, of the reasoning that underlies our consensus. It is not that we hold creationists in contempt, but the gulf that separates the scientific and the creationist ways of thinking about the world and the methods of its cognition appears too wide to be successfully bridged. Neglect is indeed the way evolutionists generally react to creationists in Europe. Why bother to engage in a serious debate with someone who is not willing to consider our arguments? The pattern of our reaction could well be the same in America, if not for the necessity of legal action in defense of evolution in school curricula. But even in America we are only forced to defend, rather than seriously substantiate, our position about the reality of biological evolution. And since attack is the best strategy of defense, we are bound to undertake a refutation of the creationist arguments instead of a presentation of our own rationale.

What we cannot ignore, however, or flatly dismiss as baseless within the worldview to which we adhere, is the challenge to the concept of evolution voiced by a group of eminent contemporary biologists—the members of the school of systematics called the pattern, or transformed, cladism—for we deal here with people who apparently share our way of thinking about the world and the methods of its cognition, with people who should therefore reach conclusions similar to ours.

In spite of their undisputable scientific and even biological background, which puts them in a sharp contrast to so-called scientific creationists, the transformed cladists—Gareth Nelson, Colin Patterson, Norman Platnick (see Bethell 1985), and their colleagues (for example, Thomson 1982, Janvier 1984)—argue against the concept of evolution as the established explanation for life's diversity. In their view, the process of genetic and phenotypic change within living populations has been abundantly documented but there is no compelling evidence for any single species having evolved from another species, let alone for the whole biosphere having originated and developed due to such transformations. The extrapolation from microevolutionary observations on particular living populations to larger temporal, spatial, and taxonomic scales may not be valid. The transformed cladists propose therefore that the concept of evolution be tested by establishing the natural systematic hierarchy of all taxa, and if such a hierarchy exists, then it would provide positive evidence for evolution.

What is meant by "natural hierarchy" in this context is that regardless of the characteristics that are taken to define a hierarchical classification, the resulting groups of objects are always the same. By contrast, an artificial hierarchy is one of several possible but mutually incongruent classifications, each of them being based on another set of characters. For example, if one undertakes to classify buttons, one can easily achieve a hierarchical classification by deciding to accept, say, the size of a button as the most important characteristic, the material it is produced from as the second most important, and the overall shape, color, and so on as decreasingly important for classification. The hierarchical classification of the same buttons, however, will be quite different if different ranks are assigned to the button characteristics. This artificiality of any hierarchical classification would be a sufficient indication of the independent origins of many kinds of button.

Evolution, by contrast, is bound to produce a natural hierarchy of species because, provided that all species originate from other species, they all arise by a branching process that automatically imposes upon all of them a consistent hierarchical structure. A discovery that the systematic hierarchy of species is indeed natural would therefore imply its evolutionary origin. In the view of transformed cladists, it is the prime task of biologists—and in particular, systematists—to determine if the hierarchy of species is natural; only then shall we be able to conclude that species really evolve from other species.

To be sure, the transformed cladists firmly believe in evolution, but they do not accept it as an empirical fact or a logical necessity. A lot of research must first be done on systematics before their belief will be replaced by certainty, and it is entirely conceivable that it will never be.

When confronted with this clash between the evolutionist and the transformed cladist ideas concerning the extent of our knowledge about the real world—despite a basic agreement on the methods of its acquisition—we are forced to spell out clearly what we feel is the ultimate justification for our position. Why, then, do we accept evolution?

At this level, it is impossible to speak for others; everyone has to speak for herself or himself. Perhaps the opinions on which we base our consensus about evolution vary among evolutionists. Hence the question is, Why do I accept the concept of evolution as established? Why do I adhere to the view that species evolve and give rise to other species, and that the whole diversity of the biosphere has arisen as the result of this process?

Insofar as one decides to stick to the worldview presumed by all science, each idea about the nature of the world can only be accepted provisionally, as the best one among those currently available on the marketplace of ideas. Whenever a new idea appears, the old ones must be reexamined in its context. The relative value of ideas is determined by testing procedures in accordance with whatever criteria one accepts, but a good test in science always reduces to a choice between at least two alternatives. Therefore, before I can attempt to present my rationale for being an evolutionist, the first step must be to present the alternatives to the concept of evolution. What is, then, the range of potential rival explanations for the existence and diversity of organic species?

First, the entire biosphere might in fact represent only one single species, in which case there would be nothing to explain, or even merely describe, by the concept of evolution. This is not a viable hypothesis, however, because it is self-evident from observing nature that the variety of life forms is enormous. And even if one chose to call

all of them one single species, it would only be a verbal trick, and the diversity of life still would call for an explanation—either by evolution, or by Creation. Second, all organic species might be created jointly—either at the beginning of the Universe, or at a later time—and simply persist since that early moment, perhaps undergo minor transformations or extinction but not evolution in the sense of giving origin to new species. The former of these two concepts of all species as Godly creations follows from the literal reading of the Bible, and it also implies that species are at least potentially eternal; the other concept is one possible reading of the biblical account of Creation as a metaphor. Individual species might also originate by means of special creation, each in its own time or perhaps in clusters. This last concept was advocated in the first half of the nineteenth century by Charles Lyell, before his conversion to evolutionism. Finally, organic species might arise spontaneously from inorganic matter. The transformed cladists must either entertain one of these alternatives, or accept willynilly evolution as the best available explanation for the existence of species; if not they would be forced to regard the origin of species as essentially incomprehensible.

The idea that all organic species were the product of a single act of Creation was refuted long before Charles Darwin. The fossil record provides clear evidence of the grand temporal succession of organic forms. Among vertebrates, for example, the sequence of appearance in sedimentary strata begins with fish, then come amphibians, then reptiles, then mammals, then hominids, and finally humans. On a lesser scale, Lyell distinguished several geological epochs, or time intervals, in the Tertiary on the basis of such a temporal succession among marine bivalves and gastropods. A similar rationale underlay also the definition of the majority of other divisions of the geological timescale. The universality of this grand pattern of the fossil record speaks sufficiently strongly against the single act of creation of all species, and this is in fact the argument that Darwin himself regarded as his most powerful one. As noted by John Maynard Smith (1986), if a fossil rabbit were discovered in Cambrian, more than 500-million-year-old strata, this would indeed be a very serious challenge to the validity of the concept of evolution.

The paleontological evidence, however, cannot refute the idea of special creation of each individual species. The fossil record itself provides no compelling evidence to substantiate the claim that new species could not possibly be created one by one, in agreement with the grand pattern of the geological succession of organic beings. For, as it was well appreciated by Darwin who never employed the paleontological evidence in this way, the fossil record contains so many morphological gaps between what might be regarded as ancestral and descendant species that it gives the impression of discontinuity rather than evolutionary continuity. Even granting the reality of some (the optimist would perhaps say, several) instances of continuity in the record of a modern species and its fossil ancestor (for example, the evolutionary history of the polar bear; see Kurtén 1964), these instances are too few and isolated to build into a strong statistical case for evolutionary transitions between species. Moreover, the naturalist's observations do not prove evolution either. For although the potential for biological change under either artificial or natural selection has been convincingly and abundantly demonstrated in various species, it shows only that species undergo evolution; this is not a compelling proof, however, that new species indeed arise by such microevolutionary processes.

The simplest argument for evolution as the mechanism of species origination was provided by Richard Lewontin (see Bethell 1985). As he put it, it is an empirical statement that all organisms have parents. It is also an empirical statement, based on the grand pattern of the geological succession of organic beings, that there was a time when no mammals existed on Earth. It is therefore only logical to conclude that mammals must have originated from some nonmammals. Logically, this transition could be a slow and gradual process of transformation, or it could be an instantaneous leap, but it must have taken place somewhere and somehow; unless, that is, mammals could arise spontaneously from muck—an idea that was definitely discarded a couple of centuries ago—or be created by God's will. This argument cannot therefore ultimately refute special creation as the mechanism of species origination. On the face of it, special creation indeed is a perfectly valid hypothesis. This is why even such avowed empiricists as Georges Cuvier and Charles Lyell could accept it.

Several arguments can in fact be marshalled against the concept of special creation, but these arguments are based on evidence that neither Lyell, nor Cuvier, nor for that matter Darwin, could know, although it certainly should be known to the modern transformed cladists. All species share some apparently universal features, and it is hard to believe that this universality could exist if the species originated independently of one another. The biochemical machinery of life is the same in all species that have thus far been investigated. Organic molecules—those building blocks of life—can exist in two mirror-image variants, and ordinary chemical reactions produce equal amounts of left- and right-handed molecules. The biochemical machinery, however, could not possibly work if organisms consisted of both right- and left-handed molecules because the crucial biochemical reactions depend upon an exact fit between the spatial structures of molecules, which would then be inachievable. In fact, the protein amino acids are all left-handed in all organic species. No species is known that would use right-handed amino acids for synthesis of its molecules, but there seems to be no thermodynamic explanation for this phenomenon; left-handed proteins are neither more stable than their right-handed equivalents, nor more capable of catalyzing biochemical reactions. It is therefore reasonable to conclude that the universal occurrence of left-handed proteins indicates a common ancestry of all species rather than a fundamental and inevitable biochemical constraint on living beings.

Perhaps even more telling is the universality of the genetic code. The meaning of particular "words" of the code is very much the same across the biosphere. The same triplets of DNA bases are translated into the same amino acids in almost all species. The empirical evidence for this claim first came from the classic experiments of Ehrenstein and Lipmann (1962) who injected the messenger RNA—the genetical decoding device—for hemoglobin of rabbit into the common bacterium *Escherichia coli,* and the bacterium reacted by producing the hemoglobin. Similar phenomena have been abundantly demonstrated by the spectacular successes of genetic engineering. The genetic code thus appears to be universal. This finding per se still may not necessarily reflect the common ancestry of all species in contrast to their independent origins, for there might in principle exist some purely chemical reasons for the universal and unequivocal relationship between the "words" of the code and their meaning. Thus far, however, no such reason has been found, and the genetic code seems to reflect a frozen accident, that is, a chance configuration inherited by living species from their remote common ancestor (Crick 1968).

Admittedly, there are structures in the cell of various organisms—mitochondria, or small membranous bodies where oxidation energy from metabolism is converted into chemical energy usable by the cell—that contain their own DNA and synthesize some of their proteins according to somewhat different versions of the genetic code

(Schweyen et al. 1983). Mitochondria, however, are widely thought to represent relics of very ancient (earlier than the frozen accident) events during which some very primitive prokaryotic organisms were incorporated as symbionts into the more advanced eukaryotic cell. More troubling is the recent discovery that some ciliate protozoans also employ a slightly different version of the code; this is most easily interpreted as a late evolutionary event in this side branch of the eukaryota (Grivell 1986). However, even if one would accept this variation in the genetic code as evidence of independent origins of a couple of the major groups of organisms, there still would remain the uncontroversial universality of the code among all higher eukaryota as well as among the vast majority of prokaryota. It is hard to believe that this might merely reflect a coincidence.

Additional arguments for evolution as the mechanism of species origination comes from systematics, the very field of the transformed cladists. For even though they are perfectly right that we cannot as yet be sure if the systematic hierarchy of organic taxa is entirely natural, there are several strong indications that it indeed is natural. This claim is supported by the congruence between systematic hierarchies based on very different characteristics. For example, Ford (1944) pioneered the use of pigment chemistry in taxonomic analysis of butterflies. His results were entirely consistent with the systematic arrangement established earlier with use of more traditional methods based on the anatomy of these forms. Penny and coworkers (1982) undertook to classify a number of species by means of analysis of the amino acid sequences in five different proteins. The classification they obtained was the same regardless of which protein they considered. At a more basic level, the remarkable stability of the main outline of biological classification from pre-Darwinian times until the present age of DNA-sequencing and genetic engineering strongly suggests the naturalness of the systematic hierarchy, because the ranking of particular characteristics has obviously undergone a considerable change in the meantime.

Certainly, systematists could not thus far, and presumably will never be able to, test the entire biosphere for naturalness of the systematic hierarchy. There are myriads of living species, let alone the extinct ones, and they have innumerable characteristics that could be used for classification. It is not inconceivable that some incongruence, perhaps even a significant one, will eventually show up in the systematic hierarchy. On the other hand, however, one cannot ignore the fact that whenever a section of the hierarchy has been tested, it always turned out to be natural. This result suggests that, contrary to the transformed cladists, the systematic hierarchy is natural. The naturalness of the systematic hierarchy, in turn, implies, in full agreement with the transformed cladist opinion, the evolutionary origins of all species.

All of these arguments for evolution as the mechanism of species origination, however, are merely inductive, that is, based on a generalization from the current knowledge. Yet although induction is an adequate way of proving mathematical theorems, it cannot provide ultimate proofs about the nature and the history of the world. A new example, or better a set of examples, may always contradict the generalization. In other words, none of these lines of evidence demonstrates (and I wonder if any of them ever can) beyond any doubt that the evolutionists' consensus about biological evolution is indeed valid. This consensus is only valid if the requirement of rational reasoning is accepted.

The word "rational," as I use it here, means that the principles of logic must be observed and that supernatural causes must not be invoked, at least as long as natural processes can adequately account for the considered phenomena. In addition, the criteria used to distinguish between natural and supernatural forces and to determine the adequacy versus inadequacy of an explanation are explicitly, though not necessarily unequivocally, specified. Rationality, then, is here understood very broadly, as a kind of a rule indicating the preferred pattern of explanation of phenomena. It implies, among other things, that ultimate certainty about the truth of statements about the nature of the world is impossible. Such rationality is in fact the hub of what I regard as the scientific worldview.

Within this worldview, evolution is among the best supported hypotheses that have ever been proposed to explain natural phenomena. The concept of evolution offers the best rational explanation for the apparent naturalness of the systematic hierarchy, for the universality of the genetic code and the biochemical structure of life, and for the grand geological succession of fossils; just as gravitation is the best available causal explanation for the movement of distant bodies in space.

Rationality, however, is only one among several possible ways to think about the world and ourselves. It is one among several possible metaphysical options. And there are no compelling reasons to regard it as the best choice. One is always entirely free to follow other paths of cognition and to accept other patterns of explaining natural phenomena. Moreover, the criteria of rationality—the line of demarcation between natural and supernatural agents, and the criteria of an explanation's adequacy—cannot be established objectively and permanently. What is rational here and now may not have been so for ancient Greeks or Romans, or even for Lyell or Darwin. This fundamentally historical nature of rationality does not imply, however, that no criteria of rationality exist. For example, supernatural causes are generally understood as forces whose action is directed toward a certain goal in the future and cannot be logically reconciled with what we know about the workings of nature. Since such criteria at least temporarily exist, what is rational for me is also rational for my fellow scientists; although there are borderline situations—hypothetical forces that may be regarded as natural, and potential explanations that may be accepted as adequate, by some but not all scientists.

It requires a leap of faith to opt for rationality. But once this leap is made, the concept of species evolving from other species appears to be the best currently available explanation for the nature and history of life's diversity. In this sense, it is established and no longer questioned in science. It is equally valid as gravitation, but also equally anchored in the option for rationality.

This conclusion must also hold for our own nature and history as a species. We fit to the grand geological succession and the natural systematic hierarchy of organic forms, and we share in the universality of the genetic code and the biochemical machinery of life. *Homo sapiens* is a part of the biosphere and must therefore have evolutionary roots. It requires, however, still another leap of faith to assert that there is nothing more to humankind than biology. Some evolutionists do make this leap, while others make a similar leap but in the opposite direction or withhold an opinion and profess agnosticism. All these options have a long and respectable tradition. Given the obvious inadequacy of all scientific explanations for the relationship between human mind and body, this should not be particularly surprising.

We do not know what the mind is and how it originated, nor do we even know how to identify its presence objectively. One may believe that the mind will eventually be fully explained by, and hence reduced to, some physical processes; or that reductionism must be altogether abandoned as a scientific research strategy if the mind is ever to be understood by science; or else that the mind should be excluded from the realm of science and referred instead to the domain of philosophy or even religion. Here stops the rationality of the biologist. The religious view of human nature is therefore perfectly compatible with evolution. This possibility has long been recognized by, for example, the Roman Catholic Church, which accepts evolution as the mechanism of the origination of human body and brain but maintains that God created and gave us the soul.

We, evolutionists, do not all agree on this point, but we do not argue about it either. We only maintain that, once rationality is adopted as the norm of our reasoning, evolution is the best available hypothesis on the mechanism of species origination and the causal force of life's diversity. It is this consensus that makes up the common ground for all real arguments about evolution. It paves the way toward an exciting realm of controversies on why and how this evolution proceeds.

Some of these controversies are the topic of this book.

BACKGROUND

2 What Do We Argue About?

One can argue about the actual course of evolution—the path it has followed from the origin of life on the Earth until today—and the methods of its reconstruction; or one can argue about the theory of evolution—the mechanisms that bring about evolution, the causal forces and the limitations they encounter.

These two major domains of evolutionary biology obviously are interdependent. The theory of evolution must be tested against empirical data derived from observations of the actual course of biological evolution. But no data can be collected and understood, no empirical knowledge can be acquired, without a conceptual, theoretical framework. In a sense, the difference between the two domains is merely in emphasis. No doubt, however, some scientists prefer to study the course of evolution, while others would rather focus on the theory. It is a matter of individual temperament more than anything else, I believe. My own preference, and hence also the focus of this book, is theoretical.

Since the 1930s and 1940s, that is, since the classic works of Ronald Fisher (1930), Sewall Wright (1931), James B. S. Haldane (1932), Theodosius Dobzhansky (1937), Julian Huxley (1942), Ernst Mayr (1942), George Gaylord Simpson (1944), Bernhard Rensch (1947), and G. Ledyard Stebbins (1050) were published, evolutionary biology has been dominated by what is now widely known as the neodarwinian, or synthetic, theory of evolution. It is an extremely successful and productive blend of the classic Darwinism with Weismannian selectionism and Mendelian genetics, enriched by subsequent advances in a host of biological disciplines—from systematics and ecology to molecular genetics and biochemistry. This blend should be more properly regarded as a paradigm rather than theory.

"Paradigm" is a word made popular in philosophy of science by Thomas Kuhn (1962) in his famous book *The Structure of Scientific Revolutions*. It was employed by Kuhn to indicate the existence of intellectual, conceptual frameworks within which all disciplines of science normally operate. A paradigm tells the scientist which ideas, concepts, and approaches are the most fundamental in his field of study; how this field relates to other disciplines of human inquiry; how to do research in this field, and what criteria to apply while judging the results of tests and experiments. Briefly, it imposes a structure on each scientific endeavor. Paradigm thus designates something less than a worldview but more than a scientific theory. Worldview is what a person believes to depict the nature of the entire world, including the human self and its relation to the outside reality. It reflects a metaphysical option, or rather a set of options that may or may not be coherent with each other. Scientific theory, in turn, is a set of formal statements about a fragment of the world, rationally explaining a class of natural phenomena by causal processes. It must be logically coherent and empirically testable by comparison of its consequences to experimental data or observations from nature.

The neodarwinian theory of evolution indeed is a paradigm, not a theory strictly speaking, for it organizes a research program in evolutionary biology and indicates a mode of explanation for evolutionary phenomena, instead of providing a strict model of evolutionary processes and identifying its domain of applicability and its boundary conditions. It is developed around a full-fledged evolutionary theory—the genetical theory of the interplay between natural selection, other genetic evolutionary forces, and the environment—that specifies a set of mechanisms of evolution in biological populations and species, and also clearly delimits the domain and the boundary conditions of their operation. This genetical theory of evolution starts with a logical deduction from the empirical fact that all living organisms are characterized by hereditary variation, multiplication, and interaction with environment which codetermines the effect of multiplication. Given these three properties of living organisms, they must be subject to the action of natural selection, that is, those organisms that more succesfully interact with environment must multiply faster than others. By this way, adaptations of organisms to environment are promoted. All organisms must also be subject to various agents affecting the nature of multiplication, and also to chance factors that influence the effects of multiplication.

The neodarwinian paradigm builds upon the genetical theory of these evolutionary forces and of their consequences for the patterns of change in all characteristics of organisms. But it is not synonymous with this theory. It is also anchored in the biochemical theory of molecular genetics, and it contains more than enough room for other biological theories relevant to evolution. Life obviously is a hierarchical phenomenon, extending from molecular structures in the cell to individual organisms to populations and species and to phenomena on a continental or even a global scale. Specific theories can potentially exist at each level of this hierarchy, the only requirement being that they be compatible with the neodarwinian paradigm. The main tenets of this paradigm are that (1) natural selection is the dominant evolutionary force responsible for changes in biology of organisms, and (2) that the genetical theory of evolutionary forces operating within populations and species in an independent physical-chemical-geological framework adequately accounts for the history of the biosphere. This is the essence of the synthetic view of evolution, as achieved by its great founding fathers (see Simpson 1953, Mayr 1963, Dobzhansky 1970, Wright 1970–1978) and other modern neodarwinians (e.g., Maynard Smith 1958, Grant 1963, Williams 1966).

The neodarwinian paradigm has always had its detractors, especially outside the Anglo-Saxon world. In France, for example, Lucien Cuénot (1941, 1951) relegated natural selection to the role of an evolutionary force capable only of leading to so-called postadaptations, or minor improvements in design of organic forms. He saw the main mechanism of evolution as a spontaneous and essentially random appearance of preadaptations, which he defined as new structural features without any adaptive significance in the organism's current ecological situation but with a potential for playing an important adaptive role in another environment. According to Cuénot, the appearance of such preadaptations is a necessary condition for evolution on any grander scale than mere changes in gene frequency in a local population.

A similar opinion was expressed by Richard Goldschmidt (1940), an emigrant from Germany to the United States, who argued for a sharp, qualitative difference between the causal processes responsible for bringing about adaptations within local populations, on the one hand, and for the evolutionary step from one species to another, on

the other. In Goldschmidt's view, adaptations to the local environment indeed result from the action of natural selection; at the species
level, however, the appearance of "hopeful monsters" is a precondition to evolution.

These or parallel views were also supported by the French zoologist Pierre-Paul Grassé (1973) and especially by the German paleontologist Otto Schindewolf (1950). Schindewolf further argued that the
evolutionary history of each organic group is largely controlled by a
set of rules analogous to those of the individual ontogenetic development. He conceived of each group as going through a "juvenile"
phase when the new body plan, characteristic of a given group, develops rapidly, virtually in a leap, from the ancestral form; then through
an "adult" phase of minor improvements and modifications by adaptation to the environment; and inevitably reaching a "gerontic" phase
when the group declines in old age and ultimately becomes extinct.
According to Schindewolf, then, the history of the biosphere reflects
the totality of such phylogenetic cycles in various groups rather than
simply the effects of the interaction between evolutionary forces and
environment.

The importance of natural selection to evolution was doubted, and
sometimes even totally denied by various scientists and thinkers who
might be labeled Lamarckians (e.g. Jones 1953, Lysenko 1956, Cannon 1959, Hardy 1965), holists (e.g. Bertalanffy 1949, Meyer-Abich
1963, Koestler 1967), vitalists (e.g., Wintrebert 1962, Vandel 1964),
and theists (e.g. Teilhard de Chardin 1955, Blandino 1960). None of
them could agree with the absence of any order or goal in evolution, as
implied by the neodarwinian paradigm for particular organic groups as
well as for the entire biosphere. On the other hand, the neodarwinian
paradigm was also strongly criticized by orthodox ultradarwinians for
admitting the operation of any evolutionary forces other than natural
selection alone (Davitashvili 1966).

Paradigms in science change, and even the criteria of rationality
change. These changes can hardly be fully described, and certainly not
fully explained, in purely rational terms. It is difficult to determine the
ultimate causes of any shift from one paradigm to another in the history
of science. To be sure, empirical or theoretical evidence against the old
paradigm does not necessarily prompt such a shift because it can always be, and historically often was, somehow accommodated within
the old paradigm—of course, with help of ad hoc, that is to say, additional and unjustified, assumptions. It is equally sure, however, that

such a paradigm shift absolutely cannot be effected without a strong empirical or theoretical case. Yet critics of the neodarwinian paradigm clearly lacked any strong arguments of this sort. Therefore, even if some of their insights were extremely illuminating—although such a claim could also be strongly contested—these critics were not part of the mainstream of evolutionary biology, and their voices attracted little attention.

In the 1980s, however, a new wave of criticisms has appeared and immediately found its way to the front pages of academic journals, scientific magazines, and popular media—for these new challenges to the neodarwinian paradigm seem to be much better founded in empirical evidence or theoretical studies. They also come from prominent evolutionists and hence are almost automatically granted a more sympathetic hearing.

There is a spate of books and articles proclaiming that a paradigm shift is currently under way in evolutionary biology. Suffice it to quote here a handful of book titles that illustrate the mood of their authors: *Beyond Neo-Darwinism,* a collection of essays edited by Mae-Wan Ho and Peter Saunders (1984); *Evolutionary Theory: Paths into the Future,* another essay collection edited by Jeffrey Pollard (1984); *Evolutionary Theory: The Unfinished Synthesis,* by Robert Reid (1985); *Unfinished Synthesis,* by Niles Eldredge (1985); *Evolution at a Crossroads,* an essay collection edited by David Depew and Bruce Weber (1985); or *Darwinism: The Refutation of a Myth* by Søren Løvtrup (1987). And these new critics are equally, or even more, vitriolic in their denunciation of Darwinism than were their antecedents. The leading advocates of this modern opposition to the neodarwinian paradigm write that it is nothing but a hindrance to the intellectual progress in biology (Reid 1985, pp. 360–361), and that it "has been put to the test and found false" (Nelson and Platnick 1984, p. 143); they declare it to be "effectively dead" (Gould 1980, p. 120), or at least inadequate (Gould 1982, Vrba 1982) and in need of being subsumed by a more comprehensive evolutionary paradigm, although the latter is outlined in a wide variety of ways (Løvtrup 1982, Barbieri 1985, Brooks and Wiley 1985, Reid 1985, Salthe 1985).

The scope of these criticisms is very broad. Some critics of the neodarwinian paradigm argue that its philosophical background is inappropriate in that it emphasizes simplification of biological phenomena through their resolution into smaller components—either subunits, or partial processes or causal chains—whereas the very nature of life

consists in complexity. In their view, evolutionary biology should therefore focus on searching for the laws determining the properties and
behavior of complex systems. There is, and there can be, no certainty
that viewing complexity as an extreme case of a simple situation will be
any more insightful or productive than its converse, the view that
simplicity is a special case of complexity. As argued by the biomathema-
ticians Michael Conrad (1983), Robert Rosen (1985), and their colleagues, it is possible within the latter conceptual framework to conceive even of physics and chemistry as subdisciplines of biology rather
than the other way around. Consequently, however, the genetical
theory of evolutionary forces might be viewed as only touching upon
some minor aspects of evolution, whereas a theory of complexity as the
real source of the driving forces of evolution would become the prime
objective of evolutionary biology.

Other critics negate the importance of the genetic evolutionary
forces, and especially the neodarwinian emphasis on natural selection,
because they believe that evolution is primarily about transformations
of the morphology of organic forms and hence, that some general laws
of such transformations may have greater effect on the actual course of
evolution than do the laws of genetics. Such laws are to be sought in the
patterns of ontogenetic development, since these patterns and the
mechanisms of their formation provide the only undeniable evidence
of morphological transformations in biology. This is the conceptual
basis for the research program launched by developmental biologists,
as for example several contributors to *Development and Evolution*
edited by Brian Goodwin and colleagues (1983). Still other critics
claim that the neodarwinian paradigm is incomplete insofar as it accepts only the genetical evolutionary forces, because they think that
the Lamarckian force of direct moulding the organism and its progeny
by various environmental agents may also operate in nature (e.g.,
Steele 1981, Mori 1986).

Perhaps the most vocal opponents of the neodarwinian paradigm,
however, are those critics who argue that although the genetical
theory of evolution is valid and although it describes at least fairly
well the causal forces driving the microevolutionary processes (that
is, those taking place within populations and species and directly
observed by biologists), these microevolutionary forces and their interplay with the environment are insufficient to account for life phenomena of a grander dimension—they are not the causes of macroevolution. In this view, macroevolution is the pattern of supraspecific

phenomena in time and space. It embraces all phenomena that concern species and even higher taxa as entities. An evolutionary trend involving several species in a lineage, linked by ancestor–descendant relationships, which evolve for a long time in the same direction, changing consistently their morphology or behavior, is a classic example of macroevolutionary phenomenon. Other examples include: evolutionary history of organic groups beyond the species level, up to the entire biosphere; changes in the proportion of various subgroups within a clade, or a group of related species derived from a single ancestral species; interrelationships, if any, between the rates of species origination and extinction and the number of species present in the biosphere; concordance of the evolutionary histories among several clades, etc. These critics of the neodarwinian paradigm, mainly paleobiologists, maintain that microevolution cannot account for macroevolution because the latter results from the action of a separate class of macroevolutionary forces. They thus reject one of the main tenets of the neodarwinian paradigm—the postulate of a causal unity of evolutionary processes on all time scales and at all levels of the systematic hierarchy. Steven Stanley's slogan, "Macroevolution is decoupled from microevolution" (Stanley 1979, p. 187), has become an epitome of this challenge to the current evolutionary paradigm and almost a battle cry in the current debates.

All of these criticisms designate major areas of disagreement and debate in modern evolutionary biology. This is not to say, however, that these are the only disputes about evolution. There is plenty of room for fervent debates even within the framework of the neodarwinian paradigm. Evolutionists argue, for example, about the roles of natural selection and chance in evolution. One of the major advances in evolutionary biology has been the neutral theory of evolution, proposed independently by the Japanese geneticist Motoo Kimura (1968) and the Americans Jack King and Thomas Jukes (1969). It postulates that much of the evolutionary change in structure of the organic macromolecules involved in the biochemical machinery of life is actually brought about by chance alone, instead of being under strict control of natural selection. It claims that a large proportion, perhaps even the majority, of point mutations occurring by copying errors in the genetic material neither damage nor benefit the organism and are therefore invisible to natural selection; whether they are established in the population or disappear from it depends then entirely on pure chance. The neutral theory explicitly refers to molecules only, whereas it leaves the domain of phenotypic change at the organismic level entirely to the action of natural selection. Thus, it remains within the neodarwinian framework. But it has nonetheless run counter to the widespread presumptions and therefore provoked an ongoing controversy among geneticists.

An even more violent controversy, with very strong social and even political undertones, has erupted among neodarwinians in response to the appearance of sociobiology in general—as outlined by E. O. Wilson (1975) and Richard Dawkins (1976)—and sociobiology of humankind in particular (Wilson 1978). Sociobiology is a research program in evolutionary and behavioral ecology based on the assumption that behavioral traits of all animals, including humans, may also have much hereditary variation and therefore are subject to the action of natural selection. Hence, it leads to analyses in purely biological terms of various patterns of animal and human behavior in a social context, in interaction with other members of the same population. Human sociobiologists have sought to explain biologically, for instance, the origins of morality and the patterns of sexuality and aggression in human societies, both primitive and modern. This research program also assumes that the individual organism—its structure, function, and behavior—which has been traditionally viewed as the object and target of natural selection, can in fact be partitioned into single traits, each of them demanding a separate evolutionary analysis. No wonder that sociobiology has provoked strongly worded rebuttals not only from philosophers, sociologists, and psychologists, who perceived it as a stance in the old "nature versus nurture" debate, but also from fellow neodarwinians.

Thus, there are many different arguments on evolution both within and without the neodarwinian paradigm. As a paleobiologist by profession, however, I focus in this book on one area of controversy—the one defined by paleobiological challenges to neodarwinism. I evaluate in some detail the claims that macroevolution is different from microevolution and that, consequently and contrary to the neodarwinian paradigm, the genetical theory of evolutionary forces operating within populations and species must be supplemented by a theory of macroevolutionary forces. In particular, I discuss four concepts that embody this claim: punctuated equilibrium, species selection, mass extinctions, and biotic diversification in the Phanerozoic.

As a first step, however, the background knowledge about the conceptual structure of the neodarwinian paradigm, the nature of paleobiological evidence, and the methodological rules of inference about evolution must be introduced. None of these three areas is uncontroversial. The reader should therefore be forewarned that my treatment of them is necessarily personal; though I hope that it will not strike anyone as outrageously deviating from mainstream opinion.

3 The Neodarwinian Paradigm

3.1 Individual Variation

All living organisms are, to use the terms introduced by David Hull (1980), both replicators and interactors. They multiply, that is, they can give rise to more than one offspring each. It is not unthinkable that some organisms do not, or did not, multiply. But if there ever existed any organisms without this intrinsic drive to produce offspring, or even if there ever existed organisms that only reproduced their numbers rather than multiplied and in fact maximized the number of their progeny, they must have been almost immediately eliminated from the biosphere by unpredictable vagaries of the environment. Multiplication is therefore ineradicable from life.

In this process of multiplication, organisms beget their like; they thus nearly replicate themselves. This is possible because all organisms have heredity. An organism is a physical body that develops in an orderly fashion from a fertilized egg—or from subdivision of the parental organism in asexual forms, as for instance in many protozoans, or from an unfertilized egg in parthenogenetic forms, as for example in many ostracodes—and then functions and interacts with its environment according to a set of rules. This physical body and the manner of its action at all levels, from molecular to behavioral, constitute the organism's phenotype. The rules according to which it develops and then functions constitute the genotype, or the genetic background of the phenotype. The phenotype results from an interplay between the genotype and the environment. Heredity means that the genetic information of an organism is transmitted in the process of multiplication to the progeny. The genotypic similarity between parental organisms and their offspring—and also between relatives that share large components of their genotypes inherited from their common ancestors, however remote in time—underlies the phenotypic similarity.

An organism's phenotype is what interacts with environment, and this interaction determines the organism's reproductive success as measured by the number of its progeny that will survive to maturity and reproduce further, or, to be more precise, by the proportion of the population its progeny will constitute in the future. The hereditary information passed on in the process of the organism's reproduction controls the phenotype; it thus affects the organism-environment interaction, which, in turn, affects the success of multiplication and hence the rate of spread of the genes, or the units of genetic information carried by the phenotype, in future generations.

A fundamental assertion of modern biology, and of the neodarwinian paradigm in particular, is that an organism's interaction with its environment does not directly affect the hereditary information to be transmitted to its progeny, or if it does, as strong irradiation may do, then its ultimate phenotypic effects in the progeny are random relative to the nature of the environment that the parental organisms and the progeny live in. The causal chain always leads from genes to phenotype to its reproductive success to gene frequency in future generations. This assertion, which was made very forcefully by August Weismann in the late nineteenth century, rejects the possibility of Lamarckian evolution through "inheritance of acquired characters," or, as Darwin himself put it, through "the effects of use and disuse." It thus leaves little room for debate about the general nature of heredity and its transmission in the process of multiplication.

This assertion is now widely accepted by biologists because it is strongly substantiated by the so-called central dogma of molecular biology, that is, by the crux of what is known about the molecular mechanisms of transmission of the genetic information during reproduction. Hereditary information is encoded in the structure of DNA molecules, which are replicated and then passed on to the progeny in reproduction. Decoding of this information during the offspring's ontogeny leads from DNA through RNA to the synthesis of appropriate proteins at appropriate times in the cell—provided of course that the biochemical compounds necessary for such synthesis are present in adequate supplies in the cell. The genetic information determines the norm of reaction of the phenotype to its environment. Hence, it codetermines the sequence and the rules of ontogenetic development and, consequently, the structure and function of the organism. Thus, the genetic background of each organism contributes to the shaping of its phenotype. Genetic information may sometimes be passed on also from RNA molecules to DNA, but it cannot be transmitted from proteins back to either RNA, or DNA. To the extent, therefore, that the nucleic acids DNA and RNA are the only carriers of hereditary information, characters acquired by an organism in the course of its ontogeny cannot be inherited by its offspring.

It is quite possible that, in addition to the nucleic acids, some other carriers of genetic information will be discovered to play a significant role in heredity. If these possible alternative carriers of genetic information would also be directionally affected by the phenotype-environment interaction, this might be a fatal blow to the neodarwinian paradigm. Some mechanisms of this sort may indeed operate in nature; for example, a culture of liver cells will continue to produce liver cells while a culture of kidney cells will continue to produce kidney cells even after a complete separation from the organism—in spite of their genotypic identity. Thus, the cytoplasmic contents and the cellular context, and not only the genotype, may affect the cell's phenotype. It is very likely, however, that they only activate or inactivate some portions of the genetic information which are present in all different cells of the organism. At the organismic level, moreover, such mechanisms of heredity do not seem at present to play any significant role.

The genetic information each organism inherits from its parents affects its interaction with environment, which, in turn, determines the relative success of its multiplication. The process of DNA replication, however, is never entirely free of errors, or mutations, and the molecular mechanism of heredity leads only to a similarity but not to the full identity of the parental and the offspring genotype. Each natural population is therefore characterized by some hereditary variation among its constituent organisms. This is particularly true of sexually reproducing organisms, in which the genotypic variation caused by mutations is supplemented and in fact overwhelmed by each fertilized egg receiving a unique combination of the genetic material, half of which is derived from the maternal genes and the other half from the paternal ones.

Much of the phenotypic variation observed in natural populations actually is nonhereditary. It only reflects the effects of environment upon the individual developmental pathway; it results directly from the organism-environment interaction. For example, whether the ontogenetic development of certain insects in the temperate climatic zone is completed in two or in three years, depends on the temperatures they encounter during their lifetime in the environment and not on the individual longevity of their parents (Bejer-Petersen 1975). The tempo of maturation and the appearance or not of the cannibalistic behavior in some ambystomatid salamanders depends primarily on the population density in particular ponds they inhabit and not on a variation in their genes (Collins and Cheek 1983). The potential to follow either one, or another developmental pathway has a genotypic, hereditary background but the choice of the actual pathway is determined phenotypically, that is, by the organism's reaction to the environmental conditions it experiences. Each organism may then more closely resemble very distantly related members of the population with which it shares the same environment than it resembles its parents, siblings, or offspring from which it is separated in time or space. This kind of phenotypic variation can certainly affect the organism-environment interaction and, consequently, the reproductive success of particular variants, but it has no direct implications for the nature of phenotypic variation in future generations.

Much of the phenotypic variation, however, is indeed hereditary, based on the underlying genotypic variation. The amount of dark pigment in the wings of the peppermoth *Biston betularia* in England, the classic case of industrial melanism (Kettlewell 1973), provides and excellent example. The resistance of humans to malaria is another classic example of this sort (Cavalli-Sforza and Bodmer 1971). The genetic information received by each organism only partly determines its phenotype. The variation in human height, for instance, has a genotypic background but it also is obviously and strongly influenced by the individual diet, that is, by the environmental conditions. As aptly noted by John Maynard Smith (1986), organisms vary because of differences in their "nature" as well as because of differences in the "nurture," but only the former differences are hereditary. All phenotypic variation affects the organism-environment interaction and, through it, the reproductive success of individuals, but in the case of variation with hereditary background—in contrast to nonhereditary variation—the effects of differential survival and reproduction of particular phenotypic variants are transmitted via the underlying genotype to later generations.

3.2 Evolutionary Forces

Logically, each population of entities that multiply and show a hereditary variation affecting their individual rates of multiplication must
evolve. The frequency distribution of such entities must change
through time. Logically, then, natural populations of organisms must
evolve. The shape of their phenotypic variation with genetic background—that is, the frequency distribution of various phenotypes and
the underlying genes—must change from generation to generation.
And if new phenotypic variants with increased probability of survival
and reproduction arise through mutations in the genetic background,
they will also increase in frequency.

This causal process leading to evolution of populations and species
by differential survival and reproduction of organisms due to their
hereditary variation is initiated and sustained by the force of natural
selection. This is not to say that all evolution is caused exclusively by
natural selection, but only that natural selection causes evolution.
Given the uncontroversial fact—empirically testable and corroborated
beyond any reasonable doubt—that organisms are characterized by
multiplication, hereditary variation, and interaction with environment
which codetermines the success of their multiplication, the evolutionary force of natural selection must operate in populations and species.
As a result, there must be selection of phenotypes with increased
probability of survival and reproduction and of the genotypes that lead
to such phenotypes, because there must be selection for hereditary
traits that contribute to the reproductive success of individual organisms. There must be selection for adaptations, that is, for such phenotypic characters of organisms which play a significant role in the
organism-environment interaction, have a genetic background, and
thus actually contribute to the organism's fitness. And there must be
selection against inadaptive phenotypic features that decrease the organism's fitness.

Fitness is a technical term referring to a probabilistic measure of
the reproductive success of individual organisms, relative to the
other members of the same population. It reflects the probability of
multiplication in future generations of a particular genotype via the
success of the phenotypes that carry it on, as compared to the probability of multiplication of other genotypes. This is the most broadly
adopted definition of fitness. Strictly speaking, however, several different definitions of fitness are in use in evolutionary biology, de-

pending primarily on the time scale on which the reproductive success is to be considered and also on whether the maximization of success or the minimization of the risk of extinction is analyzed. Whichever definition is accepted, fitness always depends on hereditary components of the individual organism-environment interaction and the ways that interaction affects the variation in survival and reproduction. It is sometimes argued—and presented as a fatal blow to neodarwinism—that fitness is unmeasurable in practice in any other way than by the actual reproductive success of individual organisms. The concept of natural selection appears then as a mere tautology, without any empirical content, because this concept states that the fittest will best spread in the population, whereas, by definition, the fittest is what best spreads in the population. This problem is more apparent than real, however. For although fitness cannot be directly measured, its distribution in a population can be roughly estimated in any given environmental context on the basis of ecology and functional morphology of the organisms. Hence, it is an empirically testable biological proposition that what seems a priori to be the fittest contributes in the long run most progeny to the population and thus increases in relative abundance.

Occurrence of a variation in fitness is a necessary and sufficient condition for the action of natural selection. The shape of this variation in natural populations gives direction to natural selection and to the resulting evolution. The higher the fitness of a genotype, the more will it contribute to the composition of future generations. There is selection for hereditary traits that help organisms in their struggle for existence and increase the probability of their reproductive success. Therefore, there is selection of phenotypes having such traits and, consequently, of genotypes that have high fitness because they bring these traits about. Thus, evolution proceeds.

The conceptual distinction between selection *of* objects or entities and selection *for* (or *against*) their traits, which was first made by Elliott Sober (1984), is very important here because selection may sometimes be accidental. Sober illustrates his idea with a very simple example. Imagine a box full of balls of various sizes. The box is subdivided by a horizontal partition with holes allowing for free passage of smaller balls but not of larger ones, and all the balls are mixed together in the upper part of the box. After a thorough shaking, all the smaller balls will be located in the bottom section, while the larger ones will remain in the upper division. In short, there will be selection for smallness causing selection of smaller balls. But if the smaller balls happen also to be green, and the larger balls red, there will also be selection of green balls, despite the absence of selection for greenness.

Such accidental sorting, as it has become fashionable to call this process of passive selection *of* without selection *for* (Vrba and Gould 1986), may have incidental effects on the phenotypic variation in natural populations—even on the variation with hereditary background— and thus lead to evolution. However, it can only operate on traits that neither considerably increase, nor significantly decrease the organism's fitness. Hence these traits can get a free ride on selection for some other traits, with which they happen to be correlated. By contrast, natural selection is the evolutionary force that causes selection of certain kinds of phenotypes, and the underlying genotypes, by selecting for their particular hereditary traits, for adaptations. The potential effectiveness of sorting in natural populations resides in the force of natural selection, of which it is a byproduct, contingent upon the existence of a genetic or developmental linkage between an adaptively neutral trait and an adaptation. Natural selection, in its turn, is an evolutionary force on its own.

Natural selection, nonetheless, is not the only evolutionary force. Mutations, or replication errors of the genetic information during reproduction, continually add to the genetic variation in natural populations. Through ontogenetic development, they also affect phenotypes, although their actual phenotypic effects may actually be dampened as well as amplified by the organism-environment interaction. Their occurrence, however, is independent of this interaction. They appear at random relative to fitness of the genotypes in a population. Since mutations, however, result from a biochemical process—replication of DNA molecules—the pattern of their occurrence is at least partly determined by the molecular structure of DNA. Some mutations are more likely than others, simply because some changes in DNA structure occur more easily, and hence more frequently, than others; some chemical bonds are more easily broken down, some components of the molecule are more easily substituted. Thus, mutation pressure arises, leading to the increased frequency of some genotypic variants and, consequently, to a change in the shape of phenotypic variation.

Chance is another evolutionary force. Fitness gives only a probabilistic measure of the reproductive success of individuals. It indicates what is most likely to happen. Its distribution in a population identifies the organisms, and therefore the genotypes, that can be expected to multiply faster than others in the population. The actual pattern of survival and reproduction, however, must also depend on vagaries of the environment. The fittest trees in a mountain forest can be burnt down by lava flow if the mountain turns out to be a volcano that resumes activity after centuries or millennia of silence. The fittest duck in a flock can be killed by the most skilled hunter in a party, while its less fit relatives will survive and reproduce. The smaller the population, the greater can be the difference between expectation and reality, because the greater can be the role of chance. In extreme cases, in very small populations, genetic drift—as the evolutionary force of chance is being called—can lead to substantial effects in evolution. It can either entirely eliminate some genotypic variants from the population, or it can fix a certain kind of gene by removing, regardless of their fitness, all the genotypes from which this gene is absent. By this way, it can wipe out certain aspects of genotypic variation and thus prevent natural selection from operating upon them.

Recently, several molecular mechanisms have been discovered to operate in natural populations; these mechanisms constitute the evolutionary force of molecular drive (Dover 1982). They result from the biochemical nature of the process of DNA replication during cell division. The action of molecular drives leads to a rapid homogenization of large amounts of the genetic material transmitted from generation to generation in a population, so that new mutations can spread in the population much faster than solely through their effects on fitness. That such a molecular evolutionary force does indeed operate in nature appears to be beyond doubt, but its actual effectiveness in evolution still is subject to hot debate among evolutionary biologists.

If the environment is constant and none of these evolutionary forces acts upon a given population, it will quickly reach the equilibrium state first described at the beginning of this century by the British mathematician Geoffrey Hardy and the German naturalist Wilhelm Weinberg. The frequencies of particular genotypic variants, and hence also the shape of the phenotypic variation, in the population will then remain constant through time. Under such conditions, there would be no evolution.

If the environment changes, however, the evolutionary forces must begin to operate. They should in fact operate even without any environmental change, for there are always mutations that cause the action of mutation pressure and perhaps also of molecular drive, and that sooner or later produce a variation in fitness; on the other hand, no population is infinite in size and thus able to evade genetic drift. And if any, or all, of the evolutionary forces comes into action, as it is always the case in nature, the Hardy-Weinberg equilibrium will not be reached; the population should evolve.

The main research program in modern evolutionary biology seeks to determine how these evolutionary forces interplay with each other as well as with the environment and the biological constitution of populations and species, and also the results of the action of these forces under particular sets of environmental and biological conditions. The genetical theory of evolution in populations and species, which lies at the core of the neodarwinian paradigm, provides tentative answers to these questions.

The key element of the neodarwinian paradigm, in fact its most fundamental claim, is that natural selection is the evolutionary force that prevails in shaping the phenotypic characters of organisms in a vast majority of situations actually encountered in nature. At the level of organisms, of their structure, function, and behavior, evolution occurs primarily through selection for adaptations. "Primarily," however, does not mean "exclusively," while the use of the word "prevail" implies that all the evolutionary forces are simultaneously at work in all natural populations.

3.3   Initial and Boundary Conditions

The action of evolutionary forces always takes place within a unique, historically determined framework of environmental and biological conditions. In the case of each particular population, the evolutionary forces operate upon some given initial conditions, predetermined by the population's past evolutionary history. The initial conditions of evolution include: the shape of the phenotypic variation, which is underlain but not fully and unequivocally determined by the nature and shape of the genotypic variation; the pattern of ontogenetic development, which translates the genotype into phenotype; and the nature of the organism-environment interaction, which depends as much on the organism as on the environment in which the population happens to live and in response to which its members have acquired specific adaptations.

These conditions are all intertwined into a complex network of feedback loops that cannot be easily partitioned or disentangled. The genotype affects the development pathway and, consequently, the phenotype, which determines the organism-environment interaction; but this interaction, in its turn, exerts a strong influence on the course of ontogenetic development and may even lead to an increase in mutation rate which contributes to a change in the genotype. The initial conditions of evolution set limits on the potential of the evolutionary forces because these forces can only transform or reshape the object of evolution—the shape of phenotypic variation—but not recreate it from scratch. As aptly observed by François Jacob (1982), evolution is a tinkerer rather than creator.

In the case of each particular population, the effectiveness of evolutionary forces is also constrained by a set of boundary conditions that determine the range of possible evolutionary changes. These boundary conditions are constituted, again, by a complex network of feedback loops involving the nature of the genotypic variation, the pattern of ontogenetic development, and the physical environment along with the organisms it contains.

There are thermodynamic stability constraints on the physical-chemical structure of DNA double helix, which is the carrier of genetic information. Therefore, not all changes in the genotype are possible. On the other hand, there are sequences of developmental events leading from the fertilized egg, or zygote, to the adult organism that cannot be cut short or skipped over, and there are developmental events that are inextricably linked to each other in the process of individual ontogeny. Therefore, not all changes in the phenotype are possible, no matter how beneficial effects on fitness they could, in theory, have for the organism in its given ecological situation. Finally, the ecological situation itself constrains the action of the evolutionary forces. For example, the availability of resources and the density of the population's natural enemies, that is, competitors, predators, and parasites, impose an upper limit on its effective size; the availability of various materials also codetermines the range of structural adaptive responses of the organism to environmental challenges; and the unavoidable action of such physical forces as gravity or electromagnetism sets stringent conditions on the form of all living beings.

Evolutionary forces always operate here and now, and it is this here and now that largely contributes to the outcome of evolution in any particular situation and at any particular level of life phenomena—from molecular structures to the form, function, and behavior of

individual organisms to the shape of phenotypic variation in populations and species. Therefore, the fundamental tenet of the neodarwinian paradigm, namely the assertion that evolution occurs primarily through selection for adaptations, is not a universal statement.

Natural selection can only work upon hereditary variation that has
a significant effect on the organism-environment interaction and, consequently, on the population's variation in fitness. If a hereditary
variation exists in natural populations that does not affect fitness—if
there is a variation with genotypic background that is adaptively neutral, so that the particular phenotypic features neither improve, nor
compromise the chances for survival and reproduction—then natural
selection must be blind to it and cannot operate upon it. Under such
conditions, genetic drift will prevail.

Motoo Kimura (1983) argues that such conditions are in fact met
for much of the evolutionary change at the molecular level. Each
protein molecule, for example, consists of a number of amino acids
linked to form a string. The sequence of amino acids in each molecule is genetically determined. If a genetic mutation occurs and a
substitution of one amino acid for another takes place, the change
will often be harmful because the molecule will significantly change
its spatial structure and lose its catalytic chemical properties. Such a
mutation will generally be eliminated by natural selection. Some
mutations may have a beneficial phenotypic effect and will therefore
spread in the population. But if a mutation is neither advantageous,
nor disadvantageous—or if it is only slightly beneficial, or slightly
deleterious—its fate will be determined by chance alone; it may be
eliminated as well as fixed in the entire population. It seems, indeed, that a large proportion of changes in protein structure are in
fact adaptively neutral, or almost neutral, and hence that molecular
evolution proceeds primarily by drift, instead of by selection.

Obviously, the more precise replication of the spatial structure of
a molecule is needed for its biological function in the organism, the
less likely is adaptive neutrality of any change in its structure. Molecular evolution can therefore proceed by drift, for instance, in those
regions of DNA molecules that are not translated into the phenotype,
but not in the regions that actually code for the biochemically active
parts of protein enzymes. In the latter ones, natural selection will
prevail.

The neodarwinian paradigm thus allows for quite a substantial
role of the evolutionary forces other than natural selection, depend-

ing on the boundary conditions of evolution in each particular case. It maintains only that selection dominates at the organismic level and in the majority of actual ecological settings. And it is the task of the genetical theory of evolution to elaborate in detail the nature of this interplay between the evolutionary forces and their biological and environmental context. In fact, much of this detail has already been described by geneticists and evolutionary ecologists.

The neodarwinian paradigm also quite naturally points out the potential limitations that these evolutionary forces may encounter. For, if developmental processes constrain in some orderly fashion the shape of the phenotypic variation subject to the action of evolutionary forces, if they constrain the possible phenotypic expression of an underlying genotypic variation, then the actual course of evolution must also be constrained. And since there are good reasons to think that not all conceivable phenotypic variants can actually be realized by developmental processes, the evolutionary forces are not omnipotent (Waddington 1957). It is the task of developmental biology to identify and explain the nature of such developmental constraints on evolution. Admittedly, however, very little is thus far really known about this aspect of the evolutionary process.

3.4 Evolution

No matter exactly how such constraints actually operate in nature, and no matter how much room natural selection leaves for the other evolutionary forces to act, the neodarwinian assertion that evolution occurs primarily through selection for adaptations has some profound implications.

Adaptations are phenotypic features that positively contribute to the organism's fitness, that is, features that enhance the organism's reproductive success through the effects they have on the organism-environment interaction. Hence, adaptations are always related to the environmental context of the evolving population; the environment includes both the physical parameters relevant for survival and reproduction and also the biological community with which the organisms interact—be it as prey or predators, competitors or symbionts. Selection is always short-sighted. It is the force that causes organisms to evolve toward increased success in coping with their current ecological situation. But except for a lucky accident, it does not bring about any safety measures for the future.

The prevalence of natural selection among the evolutionary forces implies that even though no organism is perfectly adapted to its environment, each population continually undergoes selection for adaptations to the current environment and each has a long evolutionary history of selection for adaptations to the environments its ancestors inhabited. The phenotype of each organism consists therefore largely, though not exclusively, of adaptations—past and present. Some of the past adaptations clearly are very deeply embedded in the ontogenetic developmental program, so deeply that they are virtually ineradicable; others are maintained by a constant pressure of natural selection, without which they would degenerate due to the phenotypic effects of mutations accumulated by genetic drift (Hecht and Hoffman 1986). Some other phenotypic traits arise as nonadaptations—either as simple, adaptively neutral characters that become established by genetic drift, or as byproducts, due to a developmental or genetic linkage, of selection for adaptations. Such nonadaptations may later turn out to be preadaptations, that is, features that can unexpectedly serve a significant function in a new environmental context and hence begin to contribute to the organism's fitness, thus becoming adaptations par excellence. The same may also be the fate of some past adaptations that change their function, and hence the way they contribute to fitness, in a new ecological situation.

Environment can obviously be defined by a myriad of ecological parameters, but only some of them are significant to survival and reproduction of a given population, and hence determine its effective environment. Provided a constant effective environment, both physical and biological, the force of selection for adaptations can be expected to push the modal phenotype in the population as close to the optimum as possible given the particular initial and boundary conditions of evolution. The environment, however, is never constant. Some of its components, either biotic or abiotic, always change. And even if none of the parameters of the effective environment of a particular population change, the environmental change may nevertheless affect some other organisms in the community and, through them also affect the considered population. Therefore, organisms always lag behind their environments; they always fall short, if perhaps only slightly, of their potential optimum in a given ecological situation. The further they are from the optimum, the stronger is the force of natural selection and the faster is the rate of evolution.

Every rapid change in effective environment should therefore lead to a burst of evolution. Nevertheless, even a rapid evolution should generally proceed by accumulation of small phenotypic changes rather than by large jumps, because more profound phenotypic changes are more likely to decrease rather than increase the organism's fitness. This is a statistical expectation based on probability considerations. Of course, given the vast timespans of biological evolution and the myriads of evolving populations and species, improbable evolutionary events may, perhaps even have to, happen in the history of the biosphere. Such events cannot be the rule, however.

Slower environmental changes, controlled by climatic, oceanographic, and geological processes, cause habitats to continually wander from one place on the Earth to another. Populations, therefore, tend to track the migration of their preferred habitats. More precisely, the organisms able to keep pace with their migrating habitats are more likely to survive and reproduce than are relatives left behind to cope not only with deteriorating physical conditions but usually also with new predators, superior competitors, and so on. If such tracking of the environment turns out to be, for some reasons, impossible—if, for example, the climate changes, but an animal population is trapped on an oceanic island—then the adaptations that the organisms acquired in the past may become useless or even detrimental in the new setting. They cease to be adaptations and may even be selected against. In such situations, populations undergo directional selection for adaptations to their new environmental context. They have to evolve by selection for new adaptations. If they fail, perhaps because of sheer bad luck, they become extinct. This is in fact the fate of each population—sooner or later.

A history of such directional selection may ultimately lead to disruption of the actual or potential exchange of the genetic information, which existed previously with some other populations. Such exchange of genetic information, or gene flow, among several populations occurs when occasional migrants from one population join another one, thus ensuring that the shapes of the genotypic variation are roughly the same in a group of related populations; the shapes of the phenotypic variation can differ more substantially because they are codetermined by the particular ecological situation of each population. Gene flow is the mechanism enabling several separate populations to behave as a single unit of evolution. Such populations constitute jointly a species, that is, the fundamental unit of biological evolution and, consequently, also systematics. Divergent selection in populations tracking each's particular habitat through space and time may eventually disrupt the unity of the species. The populations may undergo so much phenotypic evolution that migrants from one population will be unable to effectively join another population; reproductive isolation between the populations will develop, leading to speciation, or the origination of a new species.

Evolution, however, is not always propelled by the necessity to keep up with the pace of environmental change. Sometimes, the environment creates new opportunities. The population could then well survive even without evolving in a new direction, but the very appearance of a new avenue for evolution reshapes the population's variation in fitness and thus reorients the force of natural selection. If a subpopulation is accidentally stranded in a new habitat, a small group of deer stranded on an island, for example, it undergoes very strong selection leading it in an entirely new direction, toward some phenotypic characteristics different from those selected for in the previous environmental setting. Deer on a small island will undergo selection for smallness, because of the shortage of food and the absence of predators (Sondaar 1977). Such a stranded, isolated subpopulation is most likely to become extinct. But if it has enough luck and an adequate amount and kind of hereditary variation, then a new modal phenotype can take over due to selection for new adaptations.

Sometimes, a new resource may appear in a population's habitat—for instance, immigration of a new prey species—and some genotypes that were leading to less fit phenotypes and hence were most likely to be eliminated, may turn out to be better able to exploit this new resource. The population may then undergo disruptive selection because two different sets of hereditary phenotypic traits will be selected for as adaptations. The same effect may result from extinction of a competitor, because it will leave some resources unexploited and hence open a new evolutionary opportunity for a subpopulation that previously struggled to survive on the margin of the population's phenotypic variation. In either case, such selection may accidentally result in speciation. After some initial genetic differentiation between two incipient species, there may even arise selection for development and reinforcement of their reproductive isolation—by selecting for traits preventing their successful mating—if the intermediate phenotypes turn out to be less fit.

Thus, new species originate as an incidental effect of divergent selection for different adaptations, more often than not in different environmental settings. However, this may not be the only mechanism of speciation. Chance events and molecular processes may also lead to species origination. Speciation has long been a contentious issue within the neodarwinian paradigm, but there certainly are a plethora of mechanisms leading to the appearance of new species, depending on a wide variety of biological and environmental conditions; the multitude of possibilities and opinions is well exemplified in the recent collection of articles on speciation edited by Claudio Barigozzi (1982).

In any event, organic diversity of the biosphere increases by speciation events. During further biological evolution, the structural, functional, and behavioral divergence between species, and then groups of species, may increase, and they are usually elevated by systematists to higher taxonomic ranks—assigned to different genera, families, orders, etc. At the same time, however, populations become extinct because they fail to keep pace with environmental change, which may be physical or biological, apparently slow or obviously rapid; if all populations of a species die out, be it due to a single factor or to many independent agents, this means extinction of the species. There is nothing special about either species origination or extinction. Origination and extinction are outcomes of the same evolutionary processes that continually operate within populations and species—provided only that these processes encounter some particular configurations of physical and biological conditions. As witnessed by the history of the biosphere, such configurations are by no means extraordinarily rare.

The neodarwinian paradigm therefore asserts that the history of life at all levels—including and even beyond the level of speciation and species extinction events, embracing all macroevolutionary phenomena—is fully accounted for by the processes that operate within populations and species. There is of course a whole hierarchy of biological entities. It begins with organic macromolecules at its lower end, and it extends through cells, organisms, populations, species, genera, families, and even more inclusive clades, up to the entire biosphere at the upper end. At each of these levels, the composition and the relative abundance of various kinds of entities change

through time. For example, every systematic group of organisms produce some proteins in common with some or perhaps even all other
groups; but every group of organisms also produces some highly specific proteins it shares with no other group. If one could list all proteins and their relative abundances produced by the biosphere two
billion years ago, that list would certainly differ from the one compiled for the modern biosphere. A similar difference would be discovered between past and present frequency distributions of organisms'
structural, functional, and behavioral traits that vary between as well
as within populations and species. Similarly, the frequency distribution of specieswide or even cladewide traits that vary between but not
within species and supraspecific systematic units has substantially
changed through time. For instance, there was a time in the history of
the biosphere when no organisms yet lived on land, and there also
was a time when no animal species with skin covered with hair or
feathers yet existed. No need to argue that a fundamental change in
both these respects did indeed take place. The present-day continents
are densely populated by a variety of mammals, including ourselves.
The proportions of marine to terrestrial organisms and of scaly to
hairy and feathered vertebrates have profoundly changed over geological time. These changes also are aspects of evolution. They also
reflect a selection of biological entities.

The neodarwinian paradigm maintains, however, that the biological causal forces of this selection of and hence evolution are all located within populations and species, although the effects of their
interplay with extrinsic, environmental conditions appear also on the
supraspecific levels of biological organization. This postulate is the
main target of the antineodarwinian criticisms voiced by many modern paleobiologists. They claim that there exists a separate class
macroevolutionary forces—and hence processes—that supplement
the forces envisaged by the neodarwinian paradigm and must be considered by any comprehensive theory of evolution. To support this
claim, these paleobiologists employ evidence from the fossil record.
Therefore, the nature of the fossil record must be evaluated in some
detail before the potential existence and the actual reality of macroevolutionary forces and processes can be discussed.

Before focusing on the fossil record and its potential for settling
current debates in evolutionary biology, one final point about the
neodarwinian paradigm should be made. The genetical theory of evolution, which is at the core of this paradigm, puts emphasis on the evolu-

tionary forces. In principle, however, various constraints on the action of these forces might play an even more significant role than the forces themselves. Generally, neodarwinians rarely write about such constraints and their importance for evolution—chiefly, I believe, because very little is known about their true nature and significance.

It is conceivable that the genetic material in DNA can be organized in only a finite number of ways, thus severely limiting the genotypic variation on which the evolutionary forces operate (Kauffman 1985). It also is conceivable that some laws of form determine which structural transformations of organic forms can and which cannot be achieved in the course of biological evolution (Webster and Goodwin 1982). At both levels, general theories of constraints on evolution may, in principle, be developed, and serious effort should be made to devise such theories (Hoffman and Reif 1988). Until such full-fledged theories appear, however, and until they are demonstrated to describe a biological reality of paramount importance instead of merely what might potentially exist, their conceivability cannot be regarded as a refutation of, or even a serious threat to, the neodarwinian paradigm.

4 The Fossil Record as Data on Evolution

4.1 The Nature of Paleontological Data

The paleontological data consist of fossils derived from sedimentary rocks. Fossils are remnants of organisms that lived in the geological past. They are documents of the history of life on the Earth. Sedimentary rocks are natural accumulations of mineral particles and aggregates deposited by geological processes operating at the surface of the Earth. They are the sources of information about the ecological environments in which organisms lived in the geological past. They also are representations of geological time, because each rock sequence contains a record of events that occurred during its deposition.

The fossil record of life is notoriously incomplete. Under the conditions prevailing on the Earth's surface, dead bodies of organisms rapidly disintegrate and decay. This is due to a variety of agents and processes: bacterial action, oxidation, dissolution, mechanical wear, etc. In the course of years or decades, delicate cellular tissues are eliminated without leaving even a trace. Generally, only the hard parts of organisms have any chance of preservation in the fossil record. Hard parts are tissues either fully composed of, or at least reinforced by, some more resistant materials. These resistant materials include such minerals as carbonates, phosphates, and silicates, as well as some organic substances such as chitin, cellulose, collagen, and keratin. The resistance, and hence the fossilization potential, of organic structures made up of these various materials varies, and it also strongly depends on the specific environmental conditions in which dead organisms are placed upon death and soon thereafter. It further depends on microstructure of the hard part, that is, on the proportion of organic, easily degradable, matrix and on the spatial arrangement of more resistant particles.

Hard parts of extinct organisms are what paleontologists usually find in the fossil record. Many organic groups, however, are completely lacking any hard parts, and they can therefore be preserved as fossils only under exceptionally favorable conditions; the same holds true also for soft body parts of all skeleton-bearing organisms. The Rancho La Brea tar pit in California is a spectacular example of such exceptional preservation of large vertebrates. Other famous examples include the insects preserved in amber from the Baltic Sea area and the Dominican Republic, the amazing faunas of the Burgess Shale in British Columbia and the Mazon Creek Formation in Illinois, and the exquisitely preserved fossils of the Hunsrück Slate and the Solnhofen Limestone in West Germany. These exceptional accumulations of fossils, usually called by their German name, *Fossil-Lagerstätten,* provide paleontologists with rare glimpses of the types of soft-bodied creatures that existed millions of years ago. They offer an opportunity to more fully assess the diversity of the past life on the Earth.

Generally, however, the only remnants of organisms without hard parts to be found in the geological record are trace fossils, that is, fossilized traces of their various life activities, such as resting or foraging, which are commonly preserved in sedimentary rocks. The organisms responsible for trace fossils, however, can rarely be identified and reconstructed. Thus, a proportion of extinct organisms had virtually no chance to be preserved and must have disappeared without leaving even a slightest trace in the fossil record. It is very difficult to estimate just how large a proportion it has been. In modern marine environments, for example, it varies in a very wide range, from less than a half to the totality of a biocoenosis. As noted by Kenneth Towe (1986), a reasonable estimate is that only "10 per cent or so of the living species stand a chance of being naturally preserved in tomorrow's sediments."

Hard parts are more readily preservable as fossils, but even the most successful fossilization of a dead body or its part does not ensure that the fossil will persist in the rock and eventually fall in the hands of a paleontologist. The original materials of a fossil can be dissolved and removed, replaced, or recrystallized, and all these processes can fully obliterate the original microstructure of fossilized hard parts even though their gross appearance can be retained. Fossils can also be deformed or even completely destroyed by mechanical forces during rock deformation by tectonics or by the processes of metamorphism operating on sedimentary rocks deeper in the Earth's crust. On the other hand, they can also be excavated by erosion from older strata and reworked, that is, again subjected to the action of the same physical and chemical agents that operated on hard parts of organisms upon their death. Many fossils certainly undergo destruction as a result of such repeated cycles of burial and erosion.

All these taphonomic processes (processes leading to preservation or destruction of organic structures in sedimentary rocks) conspire to provide the paleontologist with merely a small sample of the life forms that existed on the Earth in the geological past. This sample is moreover highly biased against organisms without any hard parts and for organisms with particularly massive and resistant structures. It is also biased in terms of habitats and modes of life represented in the fossil record, because various ecological environments sharply differ in their chances of being preserved in the form of sedimentary rocks, and they also differ in the proportion of fossilizeable organisms that inhabit them. In general, shallow-water marine biocoenoses and habitats are better represented in the fossil record than either deep oceanic or continental ones.

Of course, even the best preserved fossils, with all details of their hard-part morphology, represent only a fraction of the anatomical characteristics of the organism, let alone its other phenotypic traits. Consequently, the characteristics that are diagnostic of species, genera, and sometimes families in many organic groups are inaccessible to the paleontologist. And while various groups of fossils can be distinguished on the basis of their morphological similarities and differences, their recognition for biological entities truly representative of such taxa is largely a matter of convention and intuition. Taphonomy thus very severely limits and often distorts the picture of ancient biotas available to the paleontologist.

Equally limited is the amount of information that paleontologists can acquire about the ecological context of organisms that lived in the geological past. Continents drift, mountain ranges wax and wane, climates change, sealevel drops and rises, the world ocean, the atmosphere, and the biosphere all evolve. Generally, therefore, no extinct organism lived in an environment like the one in which its fossil remains occur today. One of the prime tasks of geologists and paleontologists is to reconstruct past environments of extinct organisms in as much detail as possible. A variety of lines of evidence are usually employed to this end.

Mineral particles and aggregates that constitute sedimentary rocks originate in different ways and in different environmental settings, and are subsequently subject to the action of different geological agents in the depositional environments in which they accumulate. For instance, mineral grains hit one another as well as other objects while being transported by water or wind, but the effects of this process depend on the resistance of particular grains to mechanical wear as well as on the medium in which the process takes place. Similarly, weathering of the same mineral or rock may take different forms and lead to different end products depending on climate and water chemistry.

The composition and structure of sedimentary rocks provide clues to the nature of ancient environments in which they originated. For example, the surface of quartz grains found in deserts differs from the surfaces of those found in aquatic environments. The spatial arrangement of quartz in a rock representative of aquatic environment can indicate deposition from directional currents, wave oscillations, or turbidity currents (that is, heavy clouds of suspended sedimentary material flowing downslope along the bottom of a basin). Their size is, in general, roughly correlated with water turbulence in the environment and with proximity of the land source area of the sedimentary material; hence, the frequency distribution of their various size classes in a rock can add further information about the environmental parameters.

Petrological analysis of calcareous rocks, or limestones, usually is even more telling about their depositional environment, especially if mineralogical and geochemical evidence is also taken into consideration. The occurrence of various peculiar sedimentary structures in limestones indicates very specific environments—from supratidal flats to deep-water oceanic basins. A variety of microstructural features of calcareous rocks are characteristic of differential conditions of water agitation and salinity. The abundance relationships between calcium, magnesium, and strontium, and the stable isotope ratios of oxygen and carbon provide valuable information about water temperature, salinity, and productivity (or the richness of organic, primarily phytoplanktic, life in surface waters); other mineralogical and geochemical indicators allow for distinguishing between well and poorly oxygenated conditions.

The geographical location of ancient environments relative to the poles and the equator can often be inferred from paleomagnetic measurements. Some minerals are magnetically active and undergo magnetization by the Earth's field. Therefore, the orientation of the Earth's magnetic field in the geologic past can be derived from measurements on the rocks containing such minerals. Using paleomagnetic measurements of various rocks of various ages from all continents, geologists can plot the past positions of continents on paleogeographic maps representing different moments in the geological past. Paleomagnetic analysis thus allows for distinction between, say, depositional marine environments located on the tropical shelf and analogous habitats of the higher-latitude, cool-temperate zone.

Last but not least, fossils themselves provide much information about the nature of ancient environments—though generally only under the assumptions that the adaptive meaning of various phenotypic traits is well understood or that the life habits and environments of extinct organisms were the same as are those of their living relatives. These assumptions are neither always valid, nor always violated, and hence fossils can often, but not universally and always with caution, be employed for paleoenvironmental reconstruction. Insofar as they can be recognized as remnants of organisms belonging to specific taxonomic groups, their occurrence in a sedimentary rock indicates a certain set of environmental parameters because it identifies the critical ecological conditions that must have been met for those organisms to be able to survive. For example, the occurrence of fossil elephant tusks or antelope antlers in a rock clearly identifies its depositional environment as continental; calcareous planktic foraminifers point to the marine nature of the environment; and coral reefs provide evidence for tropical or subtropical shallow-water marine conditions. This kind of inference can in fact go beyond such crude, almost trivial, paleoenvironmental interpretations. Detailed analysis of an eighteen-million-year-old fossil assemblage found in fine-grained sedimentary rocks near Korytnica in Central Poland suggests that the original community inhabited submarine meadows of seagrass growing on the muddy buttom of a small and shallow, but normally saline, bay of a marginal sea in the warm-temperate climatic zone (Hoffman 1977). The composition of another assemblage of fossils of that age from Poland resembles, in turn, a kelp-associated community (Hoffman et al. 1978).

Of course, as many independent lines of evidence should always be considered for paleoenvironmental reconstruction as are available. Nevertheless, the results of such analyses generally are rather imprecise and uncertain because many methods are inapplicable in various circumstances and all of them are fallible. Petrological analysis of sedimentary rocks often leads to ambiguous conclusions because different geological processes may lead to the same end products. Microstructural, mineralogical, geochemical, and paleomagnetic features of a rock can all undergo substantial changes during its postdepositional history, thus rendering them meaningless insofar as the original sedimentary environment is concerned. Calcium carbonate minerals, for example, may undergo recrystallization, which destroys the original microstructure; magnetic minerals are vulnerable to remagnetization and hence their orientation may reflect the Earth's magnetic field at a later time than the moment of the rock's deposition. Moreover, it often is very difficult to determine whether the rock has actually been subject to such postdepositional, or diagenetic, transformations.

Fossils, in turn, can be misleading as paleoenvironmental indicators—even if their taxonomic position is fully and correctly determined, and if their very close relatives are known from the living biosphere—because the species they represent may have evolved and changed their ecological requirements. For entirely extinct taxonomic groups, the adaptive meaning of various phenotypic traits very often remains elusive. The potential for transportation of dead organisms or their fossils beyond the limits of their true habitats, and especially the potential for selectivity in such transportation and even destruction, can significantly obscure the ecological context of extinct organisms as it might be inferred from the associated sedimentary rocks and the co-occurring fossils. Moreover, the rate of sediment accumulation is usually so low compared to the life span of individual organisms that fossils representative of organisms that actually never lived together—and may have been separated by as little as months (under the conditions of very rapid sedimentation) or as much as millions of years (under the conditions of so-called stratigraphic condensation, due to extremely slow sedimentation alternating with periods of nondeposition)—often occur side by side in the rock, thus giving a false impression that they belonged to a single biocoenosis or community.

By their very nature, then, the paleontological data are incomplete and sometimes misleading sources of information about the history of life on the Earth. Nevertheless, the fossil record is, and must remain, one of the only two sources of information about this history. The other source is the modern biosphere, which can be analyzed by comparative-biological techniques in order to establish the genealogical relationships among various living organic groups and the sequence of their appearance in evolution. As it is the case with all historical sources, however, both these sources have their specific problems and they both must be treated with appropriate caution.

4.2 Reconstruction of Phylogeny

One of the main aspects of the history of life—and hence, one of the main challenges to the evolutionary biologist who undertakes to reconstruct and understand this history—is the origin of the enormous diversity of life forms that present themselves to our eyes whenever we trespass the limits of our artificial, urban environment. As depicted movingly by E. O. Wilson (1984) in his beautiful essay *Biophilia,* or the love of life, a deeply emotional fascination with this diversity and an insatiable need to explain it constitute the prime driving force of evolutionary research.

This diversity of life has arisen by a branching process of speciation. New species originate and diverge from preexisting ones. They subsequently give rise to more species and, consequently, also to higher taxa, which actually represent a cluster of branches of the genealogical tree of life stemming all from a single older branch. The branching is reflected in organisms by a change in DNA structure and by appearance of new phenotypic traits—be it at the level of cell biochemistry, physiological functions, developmental or behavioral patterns, or morphology of the whole organism. Distribution of such traits among taxa is therefore indicative of topology of the tree of life, which represents the genealogical, or phylogenetic, relationships among taxa.

Because of taphonomic limitations of the fossil record, distribution of phenotypic traits can best be established in the living biosphere. This is why Vincent Sarich and Allan Wilson (1967), Walter Fitch and E. Margoliash (1970), and many other molecular biologists, as well as Colin Patterson (1981, 1982), Gareth Nelson and Norman Platnick (1981), and many other adherents to the cladistic methodology of systematics, forcefully argue that comparative biology of living organisms is the only available method of acquiring a reliable knowledge about the evolutionary history of various taxa and, ultimately, about the origin of life's diversity. By analyzing the distribution of phenotypic—biochemical, developmental, morphological, etc.—traits among subdivisions of a considered taxon as well as in related taxa, it is possible to determine which traits of the investigated taxon represent more evolutionarily advanced, and which ones identify more primitive, stages in the taxon's history. Characteristics that occur both within and without a taxon are most likely to be primitive, in the sense of being inherited from the common ancestor of the taxon and its related group; whereas those traits that are progressively more confined to the taxon's subdivisions can be arranged into a branching pattern of more advanced phenotypic structures. In such a pattern, characteristics that are restricted to only one species are the most advanced, in the sense of specifying those branches of the evolutionary tree that did not yet produce offshoots; they indicate evolutionary pathways that did not bifurcate by speciation. Less advanced characteristics, in turn, specify smaller or greater clusters of branches, or evolutionary pathways that have undergone repeated splitting by successive speciation events.

A distinction between primitive and advanced traits can thus be made by the comparative-biological criterion, or the outgroup comparison criterion. It can be crosschecked with inferences made from ontogenetic patterns, on the assumption that the earlier a characteristic appears in ontogeny, the more primitive it is. In principle, the fossil record could also be employed as a supplementary criterion, because the geological succession of fossils should reflect the evolutionary sequence of taxa as defined by their phenotypic traits. Cladists, however, generally reject the application of paleontological data because of the inherent taphonomic biases that may even lead to appearance of a reversed sequence of evolutionary events.

The unreliability of paleontological data may in fact be exaggerated by cladists. There is no reason to suppose that taphonomic biases should primarily act against either ancestral, or descendant species; relative to the sequence of evolutionary branching events, taphonomy can be expected to operate at random. Statistically, therefore, the fossil record should more often than not reflect the actual sequence of events, although it may well be misleading in any particular instance (Paul 1982).

However, the criteria generally employed by cladists to distinguish between primitive and advanced traits may also be misleading. The evolution of ontogeny may often proceed by appearance of new phenotypic traits at later ontogenetic stages but it is not a regular process that would take the same route in all groups of organisms. In some instances, phylogenetically new traits appear early in ontogeny—for example, the placenta in mammals and sharks; in others, ontogenetically early traits are dropped in the course of evolution—for example, the tadpole in some frogs. Thus, the ontogenetic criterion sometimes will fail.

The criterion of outgroup comparison, on the other hand, presumes that a phenotypic trait cannot orginate independently in two or more species; this assumption is very far from being obviously true. The outgroup comparison criterion also depends on the proper choice of groups related to the taxon whose systematics is being analyzed, because the presence or absence of various traits in those groups determines which characteristics of the considered taxon should be regarded as primitive or advanced. To make a proper choice, however, requires a good deal of knowledge about systematics of the considered taxa. In many instances, the best choices may actually be extinct forms; for example, the closest relatives of pterobranch hemichordates seem to be graptolites, which underwent ultimate extinction in the Mid-Paleozoic, several hundred million years ago. Moreover, a considerable number of important extinct taxa—for instance, archaeocyathans, conodonts, tentaculites—are in fact problematic, in the sense that their phylogenetic relationships to other organic groups remain unknown (Hoffman and Nitecki 1986). Such taxa cannot be employed for outgroup comparisons, even if they actually were the most appropriate ones. More often than not, however, a distinction between primitive and advanced traits can be made for groups of living organisms, even though certainty about its correctness can hardly be achieved.

Once distribution of phenotypic traits is thus analyzed, a cladogram can be constructed for the considered taxa and their subdivisions; it graphically portrays their phylogenetic relationships by showing which biological entities are most closely related genealogically. Insofar as biochemical traits are considered, which are largely subject to molecular evolution by genetic drift (for example, amino acid sequences in a class of proteins) such a pattern of interrelationships can further be calibrated in time. The timing of particular branching events can be approximated on the assumption of roughly constant rate of evolution by drift for the investigated class of macromolecules. The potential of this kind of inference has been demonstrated by Sarich and Wilson (1967) who estimated the time of evolutionary divergence between humans and apes on the basis of immunological comparisons and concluded that the Miocene *Ramapithecus* had lived too early to be a hominid. This conclusion ran counter the consensus of paleoanthropologists of the time (Pilbeam 1972), but newer findings as well as reevaluation of older data leave little doubt that it is the molecular biologists who were right (Wolpoff 1983).

The cladogram indicates the sequence of branching events in the history of life. Thus, the pattern of evolution is reconstructed and may also be calibrated in time. Evolutionary scenarios can then be produced to explain this historical pattern and thus to provide an understanding of the history of life on the Earth (Eldredge and Cracraft 1980).

This methods works, as was demonstrated spectacularly by Stephen O'Brien and coworkers (1985) who employed analysis of a variety of molecular characteristics to establish the phylogenetic relationships between the famous giant panda and the bears, raccoons, and lesser pandas, but it is not infallible. More importantly, it can tell something about the taxa it considers but obviously nothing about those taxa that went extinct without leaving any descendants in the modern biosphere. Yet the fossil record abundantly documents the extinction of thousands and thousands of groups of organisms, many of them sufficiently different from living forms to warrant their attribution to separate families, orders, and even phyla. Therefore, the evolutionary tree of life, which is a representation of the history of life, cannot be fully reconstructed solely on the basis of living organisms. Perhaps even more importantly, however, the tree of life, or phylogeny, represents only one aspect of life's history on the Earth— a fascinating aspect, to be sure, but not the only one and even not necessarily the most interesting.

4.3 Other Questions to Ask of the Fossil Record

For many evolutionary biologists, as exemplified for instance by Richard Dawkins (1986), the main question concerning biological evolution is the problem of the origin of complex adaptations—characteristics that so extraordinarily finely tune organisms to their habitats that they were widely regarded as a proof of the existence of God, for He alone could design them so well as to astound the best engineers. This question, then, is about the universal mechanisms of evolution leading to the bewildering diversity of life forms rather than about the particular historical course this evolution has taken. The neodarwinian answer to this question points to the genetical theory of interplay between the genetic evolutionary forces, the boundary conditions constraining their operation, and the environment. But in order to test the correctness of this answer in any particular case, the enormously complex web of biological interactions must first be disentangled and analyzed in detail. Because of the inevitable taphonomic loss of information on the phenotype and the ecological context of extinct organisms—let alone on their genotype—this problem can only be addressed in studies of the living biosphere.

On a grander scale, however, one may ask if the neodarwinian answer is not only correct but also exhaustive. From the early darwinians Hilgendorf (1866) and Wladimir Kowalewski (1876), through the antidarwinians Edward Drinker Cope (1896), Henry Fairfield Osborn (1934), and Otto Schindewolf (1950), to George Gaylord Simpson (1953), a founding father of the neodarwinian paradigm, and to the latest paleobiological rebellion against this paradigm (Eldredge and Gould 1972), paleontologists have been busy studying the tempo and mode of evolution in the fossil record in order to find out if the resulting patterns are compatible with the mechanisms envisaged by evolutionary biologists. For in principle it is conceivable that although the genetic mechanisms of evolution are indeed responsible for the fine tuning of organisms to their environments, these mechanisms are nonetheless insufficient to account for the empirical patterns of morphological evolution on the time scale of millions of years—the time scale of paleontological data.

The history of life on the Earth can also be considered on a still grander scale—the scale of the entire global system, the biosphere as a whole. The main questions in this area concern the changes in biomass and diversity, in ecological and biogeographical structure, of the global biota and its various segments, and of course the causation of these changes. The feedback loops linking the biosphere with its physical-chemical environment may, in theory, be controlled by some evolutionary mechanisms beyond those responsible for the evolution of individual species; hence, one may search for mechanisms and laws of the historical process of biosphere evolution. Many scientists—from Vladimir Vernadsky (1930) and James Lovelock (1979) who argued for the global system being a single unit of evolution, on the one hand, to modern paleobiologists who focus on the internal dynamics of its evolution (Sepkoski 1978, Bambach 1983, Kitchell and Carr 1985), on the other—seem to regard this aspect of life's history as the most important one.

For many paleontologists, however, and perhaps also for the majority of laymen, the main problem about biological evolution is simply the past existence of various strange creatures that no longer live on the Earth: ammonites and belemnites, trilobites and eurypterids, blastoids and cystoids, the mysterious and bizarre *Tullimonstrum* and *Hallucigenia*, the huge mammal *Brontotherium* and the giant bird *Aepyornis*, and of course dinosaurs. How can one explain their rise and subsequent demise in evolution? The extinction of dinosaurs particularly captures the imagination. This is especially true of Americans (Hoffman and Nitecki 1985), presumably because in the American school system no sooner does a child learn to read and write than it also learns about dinosaurs and their fate. The interest in dinosaur extinction, and also in extinction of many other creatures that once inhabited the Earth, falls within a much broader domain within the history of life—a domain encompassing all major events in the evolution of the biosphere, that is, key evolutionary innovations, radiation events, mass extinctions, biogeographic restructuring of the Earth's biotas, and their impact on the pattern of life.

There are many aspects of life's history on the Earth, and in order to consider some of them—long-term dynamics of the global ecosystem, major events in the biosphere evolution, possible existence of evolutionary forces and mechanisms operating beyond the time scale envisaged by the genetical theory—the fossil record of evolution is an indispensable source of empirical information, in spite of its unavoidable deficiencies. For the fossil record to become applicable to these ends, however, a time dimension must be incorporated in the paleontological data. The historical pattern of evolution, both of life and of its physical-chemical environment, must acquire a chronology, a geological time scale. Events must be placed in a temporal sequence, and this sequence should be calibrated in real time units—be it months, years, or thousands or even millions of years—so that not only the nature of evolutionary processes can be inferred from their order in time, but also their actual tempo.

4.4   Geological Time

Sedimentary rocks—whether desert sands, lacustrine muds, or calcareous debris of a coral reef—always accumulate on a rock substrate, composed of older sediments deposited and often hardened by
lithification processes in more ancient environments. Thus, strata of
younger rocks overlay strata of older geological age. This is the law of
superposition, discovered in the seventeenth century by Nicolaus
Steno. This law can be only violated in rather unusual circumstances:
when a submarine slump, for example, takes older deposits from the
continental slope and puts them onto younger pelagic oozes in a
deep-water basin, or when powerful tectonical forces fold over huge
batches of rock strata or displace them and thrust over other areas.
The sequence of sedimentary rock strata at any locality reflects therefore the temporal sequence of depositional environments that replaced one another in that area, the sequence of local geological
events. It thus represents geological time.

By the very nature of geological processes, the time recorded by
a stratigraphic sequence at any local section is discrete, or discontinuous. Sedimentary processes are episodic. They operate momen-
tarily—bringing in and depositing smaller or greater quantities of
sedimentary material—and then cease for shorter or longer periods
of time. A centimeter-thick layer of mudstone can be deposited in a
matter of hours or thousands and thousands of years. A gap between two adjacent layers of such mudstone can represent a period
of nondeposition on the order of days or millions and millions of
years. Moreover, such gaps in the stratigraphic record often are
increased by erosion of previously deposited rock strata; for example, a turbidity current flowing downslope toward a deep-ocean basin first erodes the bottom sediments and only later lays down the
material it carries.

The completeness of rock sequences as the geological record of
time elapsed during their formation strongly varies, however, among
different depositional environments. It is greater in stable open-
ocean settings, where the sedimentation may come very close to being continuous over long time intervals, and it dramatically drops in
rocks deposited by rivers, turbidity currents, or icesheets. As judged
by comparison to modern analogues of ancient depositional environments, the differences in completeness range up to several orders of
magnitude (Sadler 1981). For past environments, however, complete-

ness of the record can only be estimated in probabilistic terms—by inference from comparison to modern environments—but not calculated with certainty for any particular rock sequence.

Neither the amount and average duration of gaps in a particular sedimentary sequence, nor even the absolute duration of the entire time interval it represents can be exactly determined. The only method of determining the absolute age, in years, of a rock is by application of the phenomenon of radioactive decay of certain chemical elements or their isotopes. Radioactive, or parent, isotopes, decay spontaneously at a known, stochastically constant rate, undergoing transformation into their daughter isotopes; for example, rubidium 87 transforms into strontium 87, potassium 40 into argon 40, uranium 238 into lead 206, and carbon 14 into nitrogen 14. It is therefore possible to calculate the original age of a rock if the current proportions of a parent and its daughter isotopes are known—provided, that is, that no daughter isotope was produced at the time of rock formation and the system has been closed ever since, no parent and daughter isotope has either escaped or been added due to diagenetic or metamorphic processes.

The precision of age determination by such radiometric techniques of course depends on validity of the latter assumption in any particular case; and it often may be violated because, for instance, some uranium-bearing minerals incorporate also various amounts of lead at the time of their formation, while argon produced by radioactive decay of potassium leaks away from crystals under slightly increased temperature and pressure. This precision is also strongly dependent on the accuracy of quantitative measurements, which obviously increases with the absolute amounts of the considered parent and daughter isotopes in a rock sample. Therefore, not all minerals are equally amenable to analysis, and not all radiometric techniques are equally applicable in all circumstances. Potassium, for example, is much more abundant on the Earth than uranium and it is therefore much more widely employed for rock dating. The amounts of the critical isotopes are also related to the rate of radioactive decay; for if it is very high, the amount of parent isotope rapidly becomes too small for accurate measurements, and if it is very low, the amount of daughter isotopes remains too small even in very old rocks. The rate of radioactive decay varies by orders of magnitude among different isotopes, and hence different radiometric techniques are employed for dating rocks of different ages: potassium–argon, rubidium–strontium, and uranium–lead for old rocks, radiocarbon solely for very young rocks. In general, the uncertainty associated with radiometric dating of older rocks is only rarely less than a few percent and can be as much as ten percent. Since the fossil record of life's evolution on the Earth extends some 650 million years into the geological past (apart from relatively rare findings of unicellular creatures in very ancient strata), the errors may indeed be substantial.

In no place, moreover, is the fossil record complete, even disregarding the erratic, episodic nature of sedimentary processes and the resulting discontinuity of all sedimentary sequences. The continental drift and the waxing and waning of seas, mountains, and icesheets have caused repeated periods of large-scale erosion. Such erosion took place essentially everywhere, though not simultaneously. As a result, the fossil record at any particular locality represents only smaller or greater scraps of the history of life—even in that particular local area, let alone beyond its narrow limits. To use the fossil record as a source of information on life's history on the Earth, one must not only determine and interpret a local sequence of geological events but also achieve a time correlation of events represented by the myriad of local sedimentary rock sequences all over the globe. To this end, radiometric techniques are inadequate. Their applicability is too narrow because of the restricted occurrence of minerals sufficiently rich in parent isotopes and sufficiently unaffected by diagenetic and metamorphic processes; and their precision is too low for time correlation of individual sequences, especially in older rocks.

4.5   Time Correlation

A variety of geological techniques are employed for time correlation of sedimentary strata, however. If a rain of small, undestructible plastic droplets fell from the sky, everywhere on the Earth at a moment in the geological past, the resulting layer of droplets would—by the law of superposition—indicate a time-parallel, or isochronous, surface in the sedimentary record of geological events. It would identify an event that took place simultaneously all over the world. If the global sealevel rose by, say, 200 meters and then dropped again by some extremely rapid, almost instantaneous, geological process, the beginning of the retreat of the sea, marked in all local sections by a characteristic sequence of sedimentary rocks, would also identify an isochronous surface. Events of these kinds are among the main tools the stratigraphers employ in their search for chronology of the history of the Earth and its life; though the actual geological events generally do not even approach the ideal time marker events given in these two abstract examples—plastic droplets do not fall from the sky, and huge marine regressions do not occur instantaneously.

Major volcanic eruptions lead, however, to instantaneous deposition, often over vast geographic areas, of ash layers that sometimes bear distinctive fingerprints in the form of geochemical and mineralogical peculiarities. Such volcanic ash layers, called bentonites if they are deposited and transformed in marine environments, can be easily identified at considerable distances and allow therefore for precise time correlation of several local sedimentary sequences. Great volcanic eruptions, however, do not occur very often—there are periods of greater and smaller volcanic activity—and volcanic ash layers do not extend over the entire globe. They cannot suffice to allow for development of a chronology of the geological and biological history of the Earth.

Some other geological events are essentially universal in scope and often very rapid. Polarity, or the positions of the northern and southern poles, of the Earth's magnetic field changes episodically and relatively very quickly, in the matter of no more than a few thousands of years. These polarity changes define isochronous surfaces in the geological record, because paleomagnetic measurements can detect the global field's polarity at the time of rock formation—provided the rock has not been subject to subsequent remagnetization—and hence identify the levels of polarity change in local sequences around the world. Rapid shifts in water chemistry of the world ocean, expressed for example by a change in stable isotope ratios of oxygen and carbon, can also be detected in the sedimentary record. Global, or eustatic, changes of the sealevel often leave very clear record in sedimentary strata and also mark isochronous geological events, although local tectonic processes may completely blur the record of such an event.

Development of a chronology based on event stratigraphy of this kind is strongly hampered, however, by the lack of individual fingerprints left by particular events. A geomagnetic polarity change and its direction—whether from normal polarity, as the current one, to reversed, or from reversed to normal—can be detected in the fossil record. But it is generally impossible to determine on the basis of paleomagnetic analysis alone which one of the scores or even hundreds of polarity changes in a given direction has thus been discovered. Similarly, a change toward lighter oxygen isotopic composition of the seawater or a reversal of sealevel shift within a global transgressive-regressive cycle can be recognized by geological analysis, but certainly scores of nearly identical phenomena of these kinds must have occurred in the history of the Earth. Time correlation of sedimentary strata by means of various geological events must therefore be aided by the biostratigraphic analysis of fossils.

Fossils, in fact, are the prime tool of stratigraphic time correlation. As early as at the end of the eighteenth century, William Smith noticed that, except for some shortcuts observed at several localities, the general order of the vertical succession of various kinds of fossils in sedimentary strata was essentially the same all over England. It follows from this purely empirical observation and the law of superposition that various fossil groups—which Smith could not even name, as he lacked an education in geology and paleontology—are characteristic of different times in the geological past, at least on a regional scale. Georges Cuvier indeed clearly saw this implication. Within a few decades, there appeared several proposals to subdivide sedimentary rock sequences according to the succession of fossils they contained. First Alcide d'Orbigny (1842–1851) and then Albert Oppel (1856–1858) established the ideal succession of strata, as determined by their fossil contents, that should occur everywhere on the Earth—had it been undisturbed by local geological history—and thus enable stratigraphers to correlate sedimentary rocks all over the world.

The concept of the geological time scale thus developed. Rock sequences are divided into erathems (Paleozoic, Mesozoic, etc.), systems (Devonian, Jurassic, etc.), and seres (Lower Devonian, Upper Jurassic, etc.), which are further subdivided into stages (Frasnian, Oxfordian, etc.), and zones (*Palmatolepis gigas* Zone, *Epipeltoceras bimammatum* Zone, etc.). Each of such divisions of strata is characterized by its fossils, be it the whole fauna, or just one fossil group, or even a single species. It thus represents a time interval during which its rocks originated and its fossils lived; since each fossil group had its own history and specific duration in time, an occurrence of its representatives uniquely specifies the geological age of the sedimentary rock.

The geological time is divided into eras (Paleozoic, Mesozoic, etc.), periods (Devonian, Jurassic, etc.), epochs (Early Devonian, Late Jurassic, etc.), ages (Frasnian, Oxfordian, etc.), and chrons

(*Palmatolepis gigas* Chron, *Epipeltoceras bimammatum* Chron, etc.); and these divisions are time intervals during which the corresponding batches of strata originated. The boundaries between such time divisions are defined by the first or last, that is, the earliest or latest, appearance of various fossils or groups of fossils and then traced from one local sedimentary sequence to another and from one geographic area to another. Their identification in distant areas means correlation in relative time, that is to say, recognition of synchronous events that occurred before or after some other events. These boundaries can also be dated by radiometric techniques and, to the extent that they can be correlated on the global scale, such datings provide an absolute calibration of the geological time scale.

The last decade has witnessed enormous progress in calibrating the geological time scale and thus establishing the absolute chronology of events in the biosphere and Earth evolution. This time scale, however, still bears very wide margins of uncertainty about the exact timing of many stratigraphic boundaries, especially prior to the Cenozoic Era. When the four most recent attempts at a calibration of the time scale (Harland et al. 1982, Odin 1982, Palmer 1983, Snelling 1985) are compared, discrepancies range up to 15 million years for several boundaries in the Middle Mesozoic and to much more than that in the Lower Paleozoic; estimates of the age of the base of the Cambrian System—which marks the onset of abundant fossil faunas with well preservable hard parts—differ by over 50 million years.

This substantial uncertainty about the true dates of geological events is only partly related to the inherent imprecision of radiometric dating techniques. At least equally important is the possibility of errors in biostratigraphic time correlation from one local sequence to another. Biostratigraphic boundaries are defined by a change in the paleontological characteristics of sedimentary strata—a particular fossil group's appearance in, or disappearance from, the fossil record—in relatively fossiliferous sections. More often than not, however, such sections do not allow for radiometric dating because they lack adequate amounts of suitable geological material; even if they contain some appropriate mineral grains, these grains may well be reworked from older rocks and hence provide wrong dates of the sedimentary rock formation. It is therefore necessary to trace biostratigraphic boundaries from one locality to another until an area is reached where they can be bracketed by reliable radiometric data. Such procedure is based on a belief that biostratigraphic boundaries indeed represent isochronous surfaces. This belief, however, is not identical with Georges Cuvier's contention that the grand geological succession of fossils reflects a temporal sequence, and it must first be justified.

The concept of the geological time scale does not depend, either historically, or logically, on the various groups of fossils that are employed for time correlation being a product of organic evolution. In fact, neither Cuvier, nor d'Orbigny, nor Oppel believed in evolution. For the purposes of biostratigraphy as they conceived of it, fossils could as well be simply various sorts of droplets having fallen on the Earth during certain time intervals. This concept could be fully accommodated by the notion of special creation of individual organic groups, from phyla down to species, and it indeed was entertained by Charles Lyell prior to his acceptance of evolution.

Thomas Huxley (1862) argued, however, that homotaxy, or similarity in the fossil contents, of sedimentary strata in distant areas does not necessarily imply their isochroneity, or identity of the times of their origination; for it might be that, for example, the Silurian faunas of North America had actually lived simultaneously with the Cambrian faunas of England, while the Silurian faunas of England represented a wave of later immigration from North America, with organisms simply tracking their preferred environments from North America to Europe. Provided that the biosphere had a similar biogeographic structure in the geological past as it has now, one has to allow for migration of ancient faunas and floras over the entire globe. If so, wrote Huxley, it is not inconceivable that the grand geological succession of fossils, which are characteristic of each group of a particular system or series, represents nothing but such a migration.

Huxley's argument can now be easily refuted on the basis of radiometric datings, which are just sufficiently precise and reliable to prove that the Cambrian faunas all over the world are indeed geologically older than the Ordovician ones, and the latter in turn are older than the Silurian. On the scale of geological periods and epochs, the observed constancy of the general order of the vertical succession of fossils provides a satisfactory justification for stratigraphic time correlation.

This is not at all obvious on finer temporal scales, however. The Jurassic System, for example, represents approximately 70 million years or so and comprises 74 standard ammonite zones (Cope et al. 1980a, 1980b). A standard ammonite zone thus approximates 1 million years in duration on the average. It certainly is beyond the resolution potential of radiometric techniques to establish the sequence of these ammonite zones in time—and even more so to determine to what extent are their boundaries isochronous—because radiometric dating errors considerably exceed the duration of particular chrons the ammonite zones represent. In a few instances, it may be possible to establish isochroneity of a biostratigraphic boundary that coincides with a very widespread layer of mineralogically or geochemically identifiable volcanic ash or with another marker of instantaneous geological events. As a general rule, however, there are no independent geological means of assessing the temporal meaning of biostratigraphic boundaries on very large distances, beyond the limits of one depositional basin.

The only certain way to achieve this end is to recognize biostratigraphic boundaries for what they should ideally represent—biological evolutionary events. The first appearance of a fossil species in sedimentary sequences should reflect its evolutionary origin, the last appearance should mark its ultimate extinction. Defined in this way, biostratigraphic boundaries are very rarely exactly isochronous over the entire range of any widespread species. Each species evolves somewhere from its ancestor and then spreads by migration over a larger area. This process of migration must take some time, although it may or may not happen very rapidly by geological standards, that is, compared to the duration of individual species and to the time represented by the smallest biostratigraphic divisions. It is also very likely that species do not become extinct instantaneously over the entire geographic range, but rather that their individual populations fall victim to various physical and biological changes in their different environmental settings and hence disappear at different times; although it is not impossible that a single agent wipes out all local populations at once.

For example, David Johnson and Catherine Nigrini (1985) used a combination of paleomagnetic correlations and radiometric datings to demonstrate that many first and last appearances of radiolarians during the last 15 million years or so were not isochronous and differed by as much as several million years between the Indian, West Pacific, and East Pacific Oceans, even within the same latitudinal zone. In their study, the last appearances were more commonly nearly isochronous than the first appearances, thus pointing to a possibility that all populations of some radiolarian species were exterminated by a single factor;

such exactly isochronous extinction of a radiolarian species was also demonstrated on the global scale by Hays and Shackleton (1976). Several examples of nonisochronous first and last appearances occur also among various benthic marine organisms (Zinsmeister and Feldmann 1984, Bretsky and Klofak 1985). The commonness of such patterns is often overlooked by paleontologists who are much too quick to regard each first or last appearance of a fossil species as defining a time-parallel surface over its entire geographic range.

Despite the certainty that biostratigraphic boundaries often are nonisochronous over large areas, however, the biological evolutionary events they represent can nevertheless be employed for time correlation of distant sedimentary sequences, because the evolutionary origination and the ultimate extinction of a species do indeed each mark very precisely a moment in geological time. Provided that no reworking from older sedimentary strata has taken place—and this can often be ascertained by geological and geochemical criteria— wherever a fossil species occurs, its presence means that the rock was deposited in a time interval between the moments of this species's origination and extinction. These two moments thus place exact limits on the geological age of the rock. By considering a large number of fossil species, it should, in principle, be possible to draw precise time correlations between remote sedimentary sequences.

Two conditions must be met, however, in order to enable the biostratigrapher to make such time correlations in the fossil record. The first condition is pragmatic. For purely practical reasons, the sedimentary sequences to be correlated must contain many fossils representative of several organic species with as short durations as possible. The shorter the species duration, the narrower the time limits its occurrence imposes on the rock's age; the more abundant the species, the greater is the likelihood of its finding in the fossil record; the greater the number of species, the easier it is to correlate from one kind of ancient depositional environment to another one, because the greater is the probability of encountering some species capable of survival under both sets of ecological parameters. Beginning with d'Orbigny and Oppel, biostratigraphers have repeatedly shown that the fossil record very often meets these requirements, although it often does not.

The second condition, however, is much more fundamental and also more difficult to fulfill. In order to achieve time correlation by bracketing the age of sedimentary strata with the first and last appearances of various fossils, it is absolutely necessary to identify the actual evolutionary origination and the true ultimate extinction of the fossil species considered. It is also imperative to identify various fossils found in distant areas as representative of the same organic species, or a set of populations acting as a single evolutionary unit due to their capacity of being linked by effective gene flow.

Obviously, there is no way to prove beyond reasonable doubt that fossils found in different localities, or even in different strata in the same locality, belonged indeed to a single species. Because of taphonomic limitations on paleontological data, hard-part morphology of organisms is the sole criterion to be applied by the paleontologist; biologists, however, have abundantly demonstrated that many species differ exclusively in behavioral, physiological, biochemical, or genetical characteristics. Many complexes of morphologically indiscernible sibling species have been discovered in the living biosphere—among groups so different as the ciliates, bivalves, insects, fish, and mammals—and their number rapidly increases with introduction of biochemical and molecular techniques of systematics. There is no reason to suppose that sibling species were less common in the geological past than they are today.

Insofar as fossil species are employed solely for the purposes of time correlation, it does not really matter if one deals with the evolutionary origin and the ultimate extinction of just one species or a complex of sibling species; provided, that is, that the moments of origination and extinction can be recognized correctly and with certainty. The real snag is that it is logically impossible to identify the moment of ultimate extinction of a species or sibling-species complex in the fossil record, because it is always possible that such fossils will subsequently be found in deposits of younger geological age. And species origination can only be recognized with certainty in the fossil record if a cladogenetic event, or a splitting of an evolutionary unit into two reproductively isolated and hence morphologically distinct ones, is actually observed in a sequence. This is not impossible but, given the incompleteness of sedimentary sequences and the commonness of speciation by divergence of allopatric populations rather than by gradual subdivision of a single population, the likelihood of finding such an example is vanishingly small; at best a handful have been demonstrated, largely among marine microfossils: foraminifers (Grabert 1959) and radiolarians (Lazarus 1986).

Theoretically at least, it should also be possible to employ as a time marker the appearance of a phenotypic trait acquired by evolving species—without a true speciation event—because that event also imposes a limit on the geological age of sedimentary strata containing fossils that already exhibit that trait. One would have to be sure, however, that this phenotypic change indeed reflects evolution, and not merely migration of contemporaneous populations tracking their preferred habitats.

Thus, even though biostratigraphic boundaries defined by biological evolutionary events should well serve the function of identifying absolutely fixed reference points for time correlation, they can rarely be applied to this end because only in exceptional cases can they be demonstrated to actually represent such events. In fact, the majority of the biostratigraphic boundaries determining the geological time scale may not represent species originations or extinctions. Generally, therefore, their exact temporal meaning cannot be clearly ascertained. It can be inferred, however, by induction from repetition of biostratigraphic patterns: a sequence that has been observed a hundred times can also be expected to occur in the next geological section.

If the same vertical sequence of biostratigraphic boundaries, or the first and last appearances of various fossil groups, recurs in a large number of geological sections representative of a variety of depositional environments over a large geographic area, the simplest explanation for such a pattern is that each individual fossil group is characteristic of a time interval and the boundaries are nearly isochronous. "Nearly isochronous" means here that the difference in exact timing of one biostratigraphic boundary between two areas is much less than difference in timing between two consecutive boundaries in each area. For, if the fossils were contemporaneous with each other and each was associated with another environment, all of them wandering over the entire area, the vertical sequence of fossils would have to be locally reversed, at least in part. Hence, insofar as the geographic scope of biostratigraphic analysis is sufficiently wide to rule out a constancy in the vertical sequence of depositional environments, the components of a recurrent sequence of fossils can be employed for time correlation.

It does not matter in this context if each fossil is a remnant of extinct organisms that lived at a specific time in a specific habitat, or if the fossils are merely physical characteristics of sedimentary rocks; it does not matter if their vertical distribution in geological sections is due to evolution or special creation or whatever other process one could think of; and it does not matter whether they can be properly interpreted in biological terms and attributed to biological species. It only matters that their vertical succession is the same in a variety of ancient depositional environments. Ironically, this is exactly the justification William Smith originally gave for biostratigraphy almost two hundred years ago.

Reasoning by induction, however, can never provide certainty about the nature of the world, and it does not lead to absolutely sure time correlations of sedimentary strata either. Biostratigraphic time correlations are always tentative, because it is always possible that the next investigated geological section will contain a reversed sequence of fossils, thus refuting the temporal meaning attributed previously to the fossil succession. Such time correlations cannot be based on individual boundaries but only on their sequences, because sequences are what can be corroborated or refuted. The greater the number of sections in which a sequence of fossils has been found to recur, the better corroborated is its usefulness for time correlation; and the more numerous the boundaries within a recurrent sequence, the more accurate the time correlations, because the more finely subdivided is the time interval equivalent to the whole sequence. There are in fact statistical methods to estimate the reliability and precision of biostratigraphic times correlation, depending on the numbers of geological sections and biostratigraphic boundaries considered (Blank and Ellis 1982).

Obviously, this reliability and precision greatly decrease at each transition from one geographic region to another, because the number of fossils common to the considered regions drops and hence the number of biostratigraphic boundaries recurring in the same vertical sequence must also drop. On the global scale, therefore, biostratigraphic correlations are neither very precise, nor very reliable. Generally, the stage level boundaries can be, somewhat optimistically, regarded as nearly isochronous in the sense given above, that is to say, relatively constant in time compared to the stage durations. It is doubtful if—apart perhaps from the Late Cenozoic—finer level biostratigraphic boundaries also are nearly isochronous in this sense.

Thus, while the fossil record is indispensable as a source of empirical information for analysis of some aspects of life's evolution on the Earth, its real value and meaning must always be very carefully evaluated in the context of the effects of taphonomic processes, incompleteness of the fossil record, imprecision and pitfalls of the biostratigraphic time correlation, and uncertainty of the radiometric dating. These uncertainties inescapably impose very severe limitations on the applicability of paleontological data to a variety of evolutionary questions. This is not to say that the fossil record of evolution is worthless, but that it must not be taken at face value.

5 Evolutionary Inference from the Fossil Record

5.1 Evolution as Explanation

The fossil record, a unique source of historical data on evolution, can bring crucial information on many aspects of the actual course of biological evolution. But can it bear also on the mechanisms of evolution? Can it provide—if only in principle—an empirical basis for the paleobiological challenges to neodarwinism, for claims about the nature of evolutionary processes? Can a historical pattern of biological phenomena offer an insight into the causal process of evolution responsible for its origination? And if it can, then how should we best proceed about inferring the evolutionary process from paleontological data? These are methodological, rather than strictly scientific, questions but they must be addressed before the arguments on the inferences actually made by various paleobiologists can be discussed and critically evaluated.

Generally, the process of biological evolution is not subject to direct observation in nature because the time spans involved are well beyond the time scale of human life. Observable are such biological phenomena as body structures, physiological reactions, and behavioral patterns of individual organisms, their interactions with the environment, and the patterns of their short-term change through time on the ontogenetic and ecologic time scales. Furthermore, the fossil record provides information about such phenomena as the patterns of long-term change in composition and distribution of various biological entities—populations, species, communities, biotas—on the evolutionary time scale. The concept of evolution as understood within the neodarwinian paradigm is an explanation for many of these phenomena.

The status of evolution as explanation in biology needs some clarification. Explanation in science is ultimately always causal. In biology, however, it may take two different forms, depending on the nature of the phenomena to be explained. In some instances, explanation of biological phenomena amounts to nothing but biophysics and biochemistry. Many physiological reactions, for example, are determined by the features of various classes of biological membranes within and between cells, and hence they are causally explained by their molecular structures. Many other biological phenomena, however, including the occurrence of one or another class of membranes in a given organism or body structure, are historical in nature. They occur here and now—or occurred at a specific place and a specific time in the geological past—and although each may share some similarities with other phenomena, each is also unique. Their causal explanation, then, is achieved by pinpointing the historical process that has actually brought these phenomena about.

To achieve this end, however, it is by far not enough to describe precisely the sequence of events that have resulted in, say, the considered structure of an organism or the analyzed pattern of temporal change in a species. A causal process, or processes, linking together in a meaningful way the events in this sequence must be identified. An answer must be provided not only to the question, how did this or that structure or pattern originate? but also to the question, why so? It is here that the concept of evolution appears. It allows for phrasing answers to the latter kind of questions. For example, an embryologist can describe the sequence of developmental stages in ontogeny of bat wings; a paleontologist can describe a shift in the average body size of a group of ammonites at successive stratigraphic levels. But it is the concept of evolution—not the embryological or paleontological description, respectively—that can explain why bats have wings or why the successive generations of the considered ammonites had progressively larger body sizes. Evolution thus elucidates and confers a meaning on historical biological phenomena.

In the context of human history, this difference between description of a sequence of historical events, on the one hand, and its explanation by a historical process, on the other, is represented by the distinction between chronicle and history (White 1965). Chronicle simply tells all that happened. History attempts to convey an under-standing of why it happened; it explains events. The mode of this explanation is fervently debated by historians because the processes considered in human history involve individual human actions, con-trolled by subjective, psychological motivations. If one wants to ex-plain, for example, the course of European history in the middle of our century, one has to consider, among other factors, the interplay between sheer madness and rational calculation in Hitler and Stalin.

Many historians and philosophers of history argue that the causes of past historical events are fully understandable, and hence explica-ble, solely in their original historical contexts which are never accessi-ble to the modern historian. A modern person can only attempt to reconstruct the situation of past historical agents, then interpret their motivations in this context, improve on this basis the original recon-struction, and proceed ever further along this path of so-called herme-neutic analysis, or reciprocal illumination, of assumptions about the context and interpretation of the motivations controlling the consid-ered actions or processes (Gadamer 1965, Ricoeur 1979). The result-ing understanding of the situation of human subjects in the past pro-vides a causal explanation for their actions and therefore for the historical process. In this view, then, historical explanation is largely intuitive.

This interpretation of the nature of causal explanation for histori-cal phenomena is rejected by many historians who believe that hu-mans generally act according to the same standards of rationality; and if so, then a knowledge of the historical context allows for logical reconstruction of the reasoning underlying decisions of past historical agents (for example, Dray 1957, Danto 1965).

This methodological controversy on human history may never be resolved because it ultimately refers to contrasting presumptions about the human nature. Natural scientists, in turn, almost univer-sally agree that they should strive for more objectivity in their causal explanations of historical phenomena. Explanation in natural sci-ences generally means that the phenomenon (or class of phenom-ena) to be explained is logically deducible from a set of general laws, or theory, taken together with particular statements specifying the initial and boundary conditions under which the processes de-scribed by this theory have operated. Explanation thus provides a set of sufficient conditions for occurrence of the considered phe-nomenon or class of phenomena; it indicates a causal process that may have resulted in the observed event or sequence of events. This is the nomologico-deductive pattern of scientific explanation (Hempel and Oppenheim 1948).

The Hempel–Oppenheim model of explanation is the one natural scientists usually undertake to apply in their research. Prerequisite to its application, however, is the existence of general theories from which to derive explanations for particular phenomena. Physical and chemical laws certainly provide a rich theoretical framework to be employed while explaining the biophysics and biochemistry of molecular and physiological reactions in organisms. It is now generally accepted by biologists and philosophers of biology that the genetical theory of natural selection and other microevolutionary forces also constitutes a set of universal laws (Ruse 1973, Hull 1974, Rosenberg 1985). Given such fundamental properties of all organisms on the Earth as their hereditary variation, multiplication, and interaction with environment which codetermines the effect of multiplication, these laws must be valid. Hence, there is absolutely no doubt that they are applicable for explanation of historical biological phenomena. It is one of the main tenets of the neodarwinian paradigm of evolution that these laws explain (that is, provide a theoretical framework to account for the origins of events) not only microevolutionary events, on the level of populations and species, but also macroevolutionary phenomena such as many patterns retrievable from paleontological observations on the fossil record.

This keystone neodarwinian assertion is rejected by many modern paleobiologists who claim that the fossil record actually demonstrates that some other biological theories must also be referred to in explanations of various macroevolutionary phenomena. This is in fact the main bone of contention in current arguments on the relationship between neodarwinism and macroevolution.

5.2 Description of Historical Biological Phenomena

A variety of macroevolutionary laws describing causal biological processes that potentially explain patterns from the fossil record are indeed conceivable. But how can we determine if one or another biological theory offers an adequate explanation for some empirical observations of historical biological phenomena? How can we decide if the neodarwinian paradigm should be modified by supplementing it with one or more macroevolutionary theories?

The first step must obviously be to identify and precisely describe the historical biological phenomena to be explained in evolutionary terms—by reference to either the genetical theory, or some other theoretical frameworks. This is not a simple task, however. Phenomena of this kind include, for example, the form of certain body structures in an organic species and their relationship to the environment, the taxonomic composition and diversity of an ancient lacustrine community or even the entire freshwater biota, the biogeographic subdivision of the world's echinoid or tetrapod fauna, and the patterns of their change through time on the appropriate time scales. To explain them, one first must identify and describe them. Difficulties and questions abound even at this initial stage of any scientific endeavor in historical biology. For, in order to be explicable by a causal evolutionary process, the recognized phenomena must represent some biological reality and not merely an artificial construct of the biologist's mind. It would be biologically nonsensical to attempt an evolutionary explanation for, say, the pattern of morphological change through time in all fossils found in a geological section, unless all these fossils are known to represent a single species, higher-rank taxon, or at least an ecological community of interacting organisms. The historical biologist must therefore be able to identify natural, functionally meaningful entities: components of a body structure, parts of a single organism, individuals representative of one species, populations interacting within the same community.

This may be a formidable challenge. Even in the case of a modern lake, it is not easy to determine objectively which birds do and which do not belong to the community. Even among living animals, it often is quite a puzzle to recognize which individuals do and which do not belong to the same species. The house mouse, which has turned out to include a number of closely related but genetically distinct, so-called sibling species (Thaler 1983), is a case in point. It is incomparably more difficult to correctly delimit the considered biological entities in the fossil record, where neither ecology, nor behavior, physiology, and genetics of organisms can be directly observed. Paleontological facts are here inextricably interwoven with the background biological and geological knowledge that may or may not be sufficiently well established to be accepted as guidelines for adequate recognition of histori- . cal biological phenomena in the fossil record. Yet those paleontological facts must somehow be extracted and arranged in an intelligible, biologically meaningful picture.

There is no single, unequivocally determined, empirical protocol to follow in order to distinguish objects for a meaningful biological study; no clearcut and certain way to identify historical biological phenomena that indeed originated by a single evolutionary process and hence can be causally explained by reference to a theory of evolution. Difficulties abound on the way to determine if a group of fossils belonged to one or more species, if they lived together or not, if they represent a biological entity or a collection of elements accidentally brought together by some taphonomic processes.

The procedure must certainly involve analysis of several independent lines of evidence and a continual attempt to explain the considered phenomenon or pattern in terms of a causal process. If a historical biological phenomenon can be plausibly explained by a process, it may really represent natural entities and their behavior rather than merely artificial constructs of the mind. Of course, finding a plausible evolutionary explanation is not the final proof that the phenomenon has indeed been identified and delimited in a biologically meaningful way, but it is at least an indication that its further analysis is worth a serious attempt.

Such initial attempts can only lead to tentative, provisional explanations, but they elucidate the identity and structure of what is out there, in nature, to be described and explained. The initial explanation of historical biological phenomena and patterns thus brings about a preliminary understanding of the underlying causal processes, which allows for improved or modified explanations, thus offering a better understanding, and so on, according to the hermeneutical principle of reciprocal illumination (Hoffman and Reif 1988).

Even a preliminary provisional but reasonable explanation, however, cannot be achieved without a sound description of the phenomenon or pattern being explained. This, again, is not a trivial task. Such descriptions should as often as possible be made in quantitative, rather than merely qualitative, terms. While describing the shell shape of, say, an ammonite species, it is not enough to observe whether it is globose or slender; for, depending on just how globose or how slender is the shell, the animal may or may not have been able to passively float in the water, and hence the shell shape would be an adaptation to very different modes of life. When the pattern of change in, for example, the size of a mammalian molar tooth upward in a geological section is being described, it is fundamental to calculate the rate of change rather than merely to notice whether the tooth increases or decreases with time; for such phenomena have very different biological meanings depending on the time spans they encompass—whether the change occurs over dozens, or thousands, or millions of years.

Descriptions of this kind naturally require a statistical evaluation of measurements taken directly on paleontological data. The measurement of shell globosity in a particular ammonite specimen and even the average globosity in a sample do not tell much about the population they came from. They give no information on the range and degree of within-population variation in this phenotypic character. In turn, if the size of the molar tooth was highly variable in each particular population of mammals in the geological sequence, the change through time inferred from differences between measurements taken on single specimens from successive samples may be more apparent than real. It may merely reflect a difference between accidentally chosen individuals representing the endmembers of the same range of variation in tooth size. Before attempting to explain such change— perhaps, though not necessarily, in evolutionary terms—the statistical null hypothesis of no change at all must first be rejected. In other words, one must estimate the probability that the observed difference is due to mere chance; and if this probability is less than a certain customarily accepted level (5% is a rule in natural sciences), pure chance is rejected as a likely cause of the phenomenon. In order to be sound, a description of historical biological phenomena or patterns may thus simply demand application of the classic statistical procedures of hypothesis testing and estimation of population parameters.

In many instances, however, such simple statistical procedures are insufficient as the sole means of accomplishing a good description of phenomena. If, for example, one can reject the null hypothesis of no change in size of the considered molar tooth, one can calculate the rate of its change through time simply by dividing the amount of change by the time span involved; alternatively, distribution of the actual data points along the time axis can be approximated by linear regression, and the slope of the line can be taken as a measure of the rate of change. Such calculations presume, however, that the analyzed data set reflects a certain orderly paleontological pattern, a linear trend, which implies a constancy of the rate of change. This presumption may well be wrong, or at least unjustified, in any given case. The data conceivably represent one or another curvilinear pattern—with the rate accelerating, slowing down, or fluctuating through time—or perhaps no regular pattern at all. Pattern must therefore be recognized in empirical data rather than arbitrarily imposed upon them by the biologist. Pattern recognition is indispensable for description of phenomena and prerequisite to their explanation.

Pattern recognition is accomplished by comparing a variety of patterns to the empirical reality and by statistical tests to determine which of them withstands the scrutiny of the data. To this end, the likelihood that a pattern's deviation from empirical data is due to chance alone is calculated for each considered pattern; the null hypothesis that the entire difference between a certain pattern and the data is due to sampling error is evaluated; and if it cannot be rejected on a given level of confidence, the pattern can be taken to adequately describe the data.

There is of course an infinite number of patterns—including a variety of irregular, random, or chaotic, ones—that could in principle be tested against the data. Models are therefore employed at this stage because they allow for picking a reasonable number of alternatives for testing. Each model of a causal process that might conceivably be at play in bringing about the considered historical biological phenomena results in a certain pattern. Thus, a number of biological models could be proposed to account for, say, the pattern of change in size of a mammalian molar tooth upwards in a geological section. For example, continual evolution by natural selection for increased grinding area of the tooth would result in a unidirectional morphological shift, most likely at a fluctuating but perhaps—given a relatively short time span and constant physical environmental conditions—at an almost constant rate. A considerable climatic change could cause migration of various populations belonging to a single but widespread species and showing variation in size of the considered tooth. Such a process would also result in a gradual and unidirectional morphological shift, but most likely at a faster rate than in the former case. A similar physical process could also affect, and cause migration of, several related but distinct species living along a temperature gradient and differing in their molar tooth size. This would lead to a pattern of more abrupt changes in the average tooth size between groups of successive samples. In turn, random sampling from highly variable but identical populations would produce a pattern of random walk, leading only by chance to a statistically significant difference between the earliest and the latest samples in the sequence. On the other hand, no plausible biological model comes to my mind that would lead to, say, a perfectly cyclic pattern of alternating intervals of exponential increase and exponential decrease in the tooth size. By comparing the patterns derived from all plausible models to the data, inadequate descriptions can be identified and rejected as statistical null hypotheses. Thus, rival descriptions of the considered phenomenon or pattern, and simultaneously also its contrasting models, are evaluated. Quantitative description of historical biological phenomena is thus inseparable from model evaluation.

5.3   The Search for an Explanation

The evaluated models do not present full explanations of the phenomena. They only outline a broad theoretical, conceptual framework and identify a family of patterns that should then be tested against the data. In each particular case, however, it remains to be seen whether the process envisaged by a certain model could indeed—under given biological and physical conditions established by independent empirical evidence—bring about the phenomenon or pattern to be explained. This is the nub of model testing in historical biology, and it generally demands much invention in finding appropriate evidence and some sophistication in statistics to fulfill this task.

Various evolutionary models often pass such tests and are therefore regarded as providing causal explanations for many historical biological phenomena. Where this is the case, the fossil record can indeed shed light on the mechanisms of evolution. More often than not, however, a rigorous testing procedure cannot be applied to models describing phenomena or patterns from the fossil record, because evidence for the relevant biological and physical conditions is unavailable. Unknown and virtually unknowable is, for example, the population size and structure of the fossil mammals showing my exemplary pattern of change in size of a molar tooth; also unknown and perhaps unknowable are many environmental factors that could affect these animals.

Explanation in historical biology thus goes beyond pattern recognition and model evaluation. Description and explanation of historical biological phenomena are nevertheless hopelessly intertwined because the models evaluated at the stage of pattern recognition always indicate at least the structure of ultimate, causal explanations for the considered phenomena. In the case of my exemplary pattern of changing tooth size, some models portray the inferred process as evolution, others as ecological replacement of populations or species, still others as merely accidental vagaries of preservation and sampling. Furthermore, the evolutionary models may construct the postulated process of evolution as propelled primarily by either genetic drift, or natural selection, etc. Models thus provide sets of general laws to be employed while seeking an explanation for phenomena from the history of life.

In a sense, each model assumes a certain simplified picture of the workings of nature. This picture consists of the general laws the model takes into account, and, obviously, no model ever considers all the laws operating in nature because it would then be as complex as the nature itself. If the pattern produced by application of such general laws to the appropriate, independently established, biological and environmental context satisfactorily fits the empirical data, an explanation of historical biological phenomenon is achieved. To consider in detail this fit is the goal of full-fledged model testing, possible solely where independent evidence is available. The way toward this ultimate step, however, leads inevitably along the path of hermeneutical analysis of descriptions shedding light on explanations and explanations allowing for better descriptions. This is so because the decision about which of the alternative pictures of the workings of nature should be used—that is, which sets of laws should be employed for testing—is already determined by model evaluation, inseparable from pattern recognition.

In particular, at that early stage of analysis it is known whether chance alone, instead of any sets of general laws, could bring about the considered phenomena or patterns. Neutral models are employed to this end (Nitecki and Hoffman 1987). Generally, neutral models are those containing as few assumptions as possible and requiring a minimum of additional theoretical justification. They most commonly represent random, or stochastic, processes because chance factors certainly operate in nature, and the assumption of pure chance as the sole cause for an empirical pattern is simpler than any alternative assumption of a combination of general laws and various initial and boundary conditions. For example, the process leading to a pattern of bird species distribution on islands in an archipelago can be modelled as chance colonization of the initially empty ecological space on archipelago by migrants from the mainland; or it can be modelled as colonization and subsequent arrangement of bird faunules according to a set of assembly rules reflecting the application of the laws of interspecific competition and natural selection to the local conditions. The former model is neutral, the latter is not because it requires justifying the relevance of these general laws in the considered context. Similarly, the process responsible for the origin of my exemplary pattern of change in molar tooth size can be represented by a stochastic neutral model of random walk, which would reflect nothing but random sampling from a highly variable population; or it can be modelled by a variety of ecological and evolutionary processes. The latter models, however, demand at least a consideration of the time scale to justify the application of particular ecological or evolutionary laws. In a sense, then, neutral models indicate the simplest explanations possible for empirical patterns under study. If a stochastic neutral model adequately accounts for empirical data, no theoretical, conceptual framework is needed to explain the phenomenon or pattern being explained.

All stochastic processes depend on chance events; hence, every run of a stochastic model will produce a somewhat different result, even when all parameters are held constant. Therefore, each stochastic neutral model provides a distinctive family of patterns rather than one unique pattern. The models are then evaluated by comparing these patterns, or some statistically established modalities of the whole families of patterns, to empirical data. Their evaluation, however, is not very different from the procedure employed for other models; it boils down to testing the null hypothesis that the degree of their misfit to empirical data falls well within the range of expectations.

A perfectly neutral stochastic model is, in fact, impossible because every stochastic model must also include a number of additional assumptions specifying the mode of operation of chance; at the very least, the probability distribution of alternative outcomes of a random process must be given. Hence, several approximately neutral models are conceivable in such particular case as representation of the processes responsible for origination of any particular historical biological phenomenon or pattern. This situation is obvious in neutral models in island biogeography, where for a decade a hot debate has raged over whether a set of species assembly rules or chance colonization is the more adequate model of species distributions on islands in archipelagoes (Strong et al. 1984). Among the main issues in this debate is the question, what exactly is meant by chance colonization? or, which approximately neutral model should be tested against empirical data?

For it indeed matters very much if the mainland, where the colonizing species come from, is regarded as a single and constant pool of species, or if it is rather subdivided into a number of different source areas; if all species are taken to have the same potential for migration, or if they rather show some differentiation in this respect, etc. In each case, the resulting pattern of chance colonization will be different.

5.4   The Dilemma of Multiple Explanations

Explanation of historical biological phenomena is thus approached through recognition of the empirical pattern and evaluation and testing of the widest possible spectrum of plausible models, including neutral ones. Whichever model passes the test of empirical data can be taken as an explanation. It may happen, however, that more than one, perhaps even several, models pass the test; or, if a rigorous test is impossible all models may appear to be plausible. The problem then immediately arises, how can we decide which one of such models should be accepted, if only tentatively, as the best explanation for the pattern? Such decisions must be made if we are to attempt an evolutionary inference from the fossil record, but with what criteria can we solve such a dilemma?

At least a part of the problem stems from the methodological reliance of modern science (including such historical natural sciences as evolutionary paleontology) on statistical testing of null hypotheses as the main tool of the scientific search for explanations. This methodology goes back at least to Karl Popper (1959) and his falsificationism. It decrees that the scientific method demands that conjectures about the nature of the world be continually tested and subject to attempts at their refutation. It starts with establishing a fundamental, logical asymmetry between verification and falsification of hypotheses. A hypothesis can be disproven but not proven. Ultimate verification is, even in principle, impossible because all consequences of a hypothesis can never be identified, articulated, and put to test. Yet, if only some predictions are correct, it is still perfectly plausible that other consequences of the considered hypothesis will contradict empirical reality, which would show the hypothesis to be wrong. Hypotheses can thus be corroborated, or strengthened, by testing ever more rigorously an ever increasing number of their consequences. Ultimate falsification, in turn, is logically possible because if only one single but inevitable consequence of a hypothesis is false—that is, contradicted by empirical data—the hypothesis itself is wrong. In principle at least, hypotheses can thus be not only undermined, or weakened, but also decisively refuted.

Popper's falsificationism rests on the belief that this logical asymmetry between verification and falsification translates also into methodological asymmetry. It presumes that this asymmetry exists not only in principle but also in scientific practice. The logical asymmetry between verification and falsification, however, may only justify the reliance on falsification as the scientific method if there exist some rules or criteria that allow for a clear and unequivocal demonstration that a consequence of the hypothesis being tested is indeed false. In other words, falsificationism as the methodology of science is contingent upon the existence of *experimentum crucis*, that is, an empirical or thought experiment, or observation, that would determine beyond reasonable doubt whether a given prediction is false.

The premise, however, that such crucial experiments exist and can be devised in practice appears to be invalid (Amsterdamski 1983). No hypothesis can ever be tested in isolation, extracted entirely from the context of other hypotheses or theories about the nature of the world. Therefore, a disagreement between hypothesis and the corresponding empirical observations does not necessarily falsify, or refute, the hypothesis, because some other elements of the plexus of hypothetical statements being compared to empirical data may be false. It is always possible that although the hypothesis is correct, it nevertheless results in a false prediction because the context in which it is tested contains one or more false elements. Consider, again, the example of a pattern of change in a mammalian molar tooth size upwards in a geological section. The hypothesis that this pattern has been brought about by continual evolution of a single lineage of populations leads to the prediction of continuity in the pattern. If the empirical pattern is discrete, however, this may imply either that the causal process was not continual evolution but rather immigration of distinct, though perhaps closely related species; or that the molar tooth size was subject to some developmental constraints that subdivided its range of variation into domains separated by threshold values; or that the sedimentation was discontinuous and the resulting fossil record contains substantial gaps.

The French physicist and philosopher Pierre Duhem (1954) postulates in fact that empirical observations can never disprove a hypothe-

sis. If an experiment or observation is in disagreement with a theoretical prediction, such a result indicates that at least one of the hypothesis
or theories involved in the plexus of processes and conditions determining the considered phenomenon is false. This corollary, however, does
not identify the one that is wrong and should be modified or replaced.

The most common counterargument to the Duhem thesis is that
when two rival hypotheses are tested in the context of the same
associated hypotheses or theories, the crucial experiment exists and
can in fact be performed because the identity of the context allows
then for recognition of the incorrect hypothesis (Popper 1963). This
argument, however, would be valid only if all the associated hypotheses were known to be correct. If they are not, a contradiction between
one of the rival hypotheses on test and the empirical observations
implies only that one of these hypotheses is wrong; but it may well be
the case that the associated context contains some false hypothetical
statements, and it may then be that it is the hypothesis contradicting
the observations which is correct rather than the one in apparent
agreement with reality. Yet it is of course very unlikely for any particular problem situation that the entire associated context of the
hypothesis being tested is known to be correct.

A stronger version of the Duhem thesis has been presented by
Quine (1980), who postulates that negative results of *experimentum
crucis* can always be reconciled with the hypothesis on test. This can
be achieved by introducing one or more additional, ad hoc hypotheses. The Duhem–Quine thesis thus states that a contradiction between predictions and observations never identifies the false hypothesis within the whole plexus of assertions about the processes and
conditions determining the considered phenomenon. Moreover, such
a contradiction does not even demonstrate that a hypothesis in this
plexus is wrong, for it can be removed by additional arguments.
Obviously, if this thesis is correct, ultimate falsification of any hypothesis is plainly impossible. No null hypothesis can then be rejected
in science.

The Duhem–Quine thesis has not been demonstrated to be univer-
sally valid, but neither has it been disproven. Stefan Amsterdamski
(1983) points out that the minimum condition that must be met in
order to refute this claim is that criteria be given that unequivocally
delimit ad hoc hypotheses as a class. Such criteria, however, do not
exist (Hempel 1966, Grünbaum 1976). What certain experts in a
scientific field or discipline regard as unacceptable ad hoc hypotheses,

others treat as well founded and reasonable auxiliary arguments. A hypothesis can then, at least in principle, be defended endlessly. Therefore, statistical testing of null hypotheses is ultimately impractical as the sole methodology of science, because it may lead to dilemmas with more than one alternative model passing the test of empirical data.

There is, in fact, no a priori, logical reason why the testing of null hypotheses should be relied upon as the prime method of pattern recognition and model evaluation. Within Popper's conceptual framework of falsificationism, and even in the more sophisticated falsificationism of his pupil and opponent Imre Lakatos (1970), null hypothesis always has a privileged position; provided that it withstands the test, it is at least tentatively accepted even though myriads of other null hypotheses are conceivable and may also pass the test. It clearly follows from the Duhem–Quine thesis—and actually even from its weaker version, the Duhem thesis, alone—that falsification of a hypothesis is in practice no more certain than its verification. The logical asymmetry between these two opposing procedures does not translate into a methodological one.

Some other methodological approaches and criteria must therefore be employed in order to determine which of several rival models should be provisionally accepted as an explanation for empirical phenomena or patterns (Hoffman 1987). One such approach, perhaps the most commonly applied one, refers to the concept of likelihood (Hacking 1965, Edwards 1972). The likelihood of a hypothesis indicates how well it is supported by a given set of empirical observations; it is measured by the fit between prediction and empirical data. A statistical comparison of the fits of rival hypotheses to the data thus indicates the best-supported hypothesis. It can then be decreed as the methodological rule of choice between competing models that the hypothesis with the highest likelihood relative to all the available data should be accepted. What is meant here is, of course, only a provisional acceptance, for another hypothesis with still more likelihood may be later proposed, or another data set may be found and change the fit of various hypotheses to the data.

For example, if a sample of ten balls is drawn at random from an urn, and they all happen to be black, the hypothesis that 90% of all the balls in the urn are black offers a better fit to the empirical data—that is, our random sample of ten balls—than the hypothesis that only 30% of the balls is the urn are black. The former hypothesis is thus better supported by the data than the latter. The hypothesis that all the balls in the urn are black, however, gives the best fit. On the strength of the likelihood criterion, then, this last hypothesis should be provisionally accepted. The qualification "provisionally" is indeed very important, for if the next random sample of ten balls will include eight black and two white balls, the 90%-black hypothesis will turn out to be most strongly supported by the data.

The likelihood criterion is questionable in principle as the methodological norm of the choice between rival hypotheses, because it consistently prefers ad hoc hypotheses, each devised specifically to fit a particular set of empirical observations, rather than more general hypotheses with their unavoidably less precise fit to the data. For instance, even if I knew that, as a general rule, the company producing urns with balls does not sell urns with all balls identical in color, I am still forced by my first sample of ten balls to accept the all-black hypothesis as the maximum likelihood one. All I can do to reconcile it with my background knowledge about the company rules, and perhaps even with a formal warranty that the balls vary in color, is to buttress the all-black hypothesis with an ad hoc one suggesting a human mistake or even a deliberate joke.

The usefulness of the likelihood criterion is also questionable in many practical situations, for if the empirical observations incorporate a large but unknown error component, the statistical comparison between rival hypotheses may well amount to nothing but measurement of their fit to statistical noise. If, for example, the color of the balls drawn from the urn were determined by a color-blind person but I were so incredibly suspicious of the urn manufacturer that I would prefer to rely more on this determination than on the company warranty, the likelihood would certainly lead me astray if the colors were indeed misidentified. Of course, the larger the error component of empirical data, the more acute is this practical problem with application of the likelihood criterion. This is precisely the case with evolutionary inference from the fossil record, since all paleontological data are very severely biased by a variety of taphonomic processes.

Therefore, still another approach is needed to address the problem in choice between various models competing as explanations for historical biological phenomena and patterns. The approach that refers to the concept of parsimony, or simplicity, may be particularly appropriate to this end. In principle, there is of course no reason to think that the simplest hypothesis is necessarily correct, or even that it comes closest to being true (Sober 1975). It is merely a matter of convention that either makes parsimony an inherent component of rationality, or accepts it as a part and parcel of the scientific method in spite of its irrationality. Regardless of which of these two interpretations is taken, the criterion of parsimony can be adopted as a methodological rule. It only must be formalized in such a way as to indicate which of rival hypotheses is the simplest, and thus to provide a clear and unequivocal solution in any particular problem situation. This may not be easy, however. Consider, for example, a random sampling of the balls from an urn. Is the hypothesis of successive drawings at random from diminishing pool of balls—those drawn are then removed from the urn—simpler as a historical explanation for a certain pattern of balls than the hypothesis of successive drawings at random from a constant pool of balls?

In fact, it is the concept of simplicity that constitutes the ultimate basis for the common, though by far not universal, preference of stochastic neutral models rather than other causal explanations for historical biological phenomena. As an explanation, chance per se does not seem to have any inherent logical priority relative to deterministic causality. It only appears to represent a simpler, more parsimonious explanation because it refers to only one agent, whereas deterministic causation generally involves several processes and their initial and boundary conditions.

Such a reference to simplicity, however, is valid only if chance is interpreted ontologically, as the very essence of the world, the sole causal agent. Stochastic models are then definitely simpler than any deterministic ones. This is clearly not the case in evolutionary biology and paleontology where randomness can only be interpreted operationally, as the net result of a huge variety of multifaceted, independent processes, often additionally obscured by enormous observational errors (Schopf 1979, Hoffman 1981). Under such conditions, it is very far from being obvious that a single deterministic process must be less parsimonious than a stochastic neutral model presuming a disorderly interaction of myriads of processes. Therefore, the criterion of parsimony alone does not even definitely solve the dilemma of whether stochastic neutral models should be preferred to deterministic causal explanations of historical biological phenomena.

To solve this dilemma, parsimony must be considered at a higher level of abstraction. No doubt, the genetical microevolutionary forces do, and must, operate in nature. This corollary follows from pure deduction based on a few fundamental properties of living organisms (Van Valen 1982). Hence, whenever a choice is to be made between hypotheses aimed at evolutionary explanation of historical biological patterns each evolutionary model must include a reference to the laws of microevolution in order to be plausible. Each model may also refer to some other biological laws, and thus go beyond the neodarwinian paradigm, but—according to the simplicity criterion—this is permissible only where an actual need to do so is demonstrated, that is, where some empirical data exist that cannot be satisfactorily accounted for by models based solely on microevolution. In general, if a set of causal processes are known to operate in nature, scientific explanations may involve other, hypothetical processes solely in situations where the known processes do not suffice to construct a satisfactory explanation under realistic (ideally, established on independent evidence) initial and boundary conditions.

This is the principle of pragmatic reductionism (Hoffman 1983). It is "reductionist" because it is a conservative principle tending to keep to a minimum the number of general laws incorporated in the conceptual framework of evolutionary explanations. And it is "pragmatic" because it only indicates a methodological rule of thumb—a version of Occam's Razor—used in choosing between rival hypotheses intended as evolutionary explanations. It is not a stance in the longstanding and never-ending metaphysical debate between holism and reductionism, but only a criterion to employ in particular problem situations, where more than one model withstands the test of empirical data. It is pragmatic also in the sense of being applicable in practice and leading to tentative rather than ultimate explanations.

The pragmatic reductionist criterion coincides in fact also with the criterion of theoretical richness, which refers to the width of the domain of applicability of competing hypotheses or models. If several hypotheses withstand the test, then, according to the latter criterion, the hypothesis should be accepted that correctly, or at least satisfactorily, predicts empirical phenomena from the widest realm of reality. In other words, the more generally applicable is a model, the more preferable it is when compared to other models. As very aptly put by Lawrence Slobodkin (1987), the criterion of theoretical richness can well be explained by analogy from everyday life. Slobodkin makes a distinction between models that are tools and models that are only gadgets. A grape peel is a gadget that can be used to peel grapes, but it can do this job well. A kitchen knife, by contrast, is a tool that can be employed for a variety of purposes, peeling grapes among many other things, although it will not do the job efficiently. While equipping my kitchen, however, I would certainly prefer to buy a kitchen knife at first and a grape peel at the end—or not at all. Some models are like grape peels, applicable to explain only a single data set; others are like good tools, pertaining to a very wide variety of empirical phenomena. According to the criterion of theoretical richness, the latter models are preferred; and they also are preferred by the principle of pragmatic reductionism, because the more general models are accepted as explanations, the smaller is the number of different models necessary to explain the world or its segment.

On these methodological criteria, then, the laws of microevolution should be provisionally accepted as the conceptual framework of evolutionary explanations wherever such explanations withstand the test of empirical data. Only if they are insufficient to account for some historical biological patterns or phenomena should evolutionary explanations go beyond the neodarwinian paradigm and invoke some other, perhaps macroevolutionary, processes.

There is no *a priori* reason to believe that the fossil record will never provide any data that would force evolutionists to supplement the neodarwinian paradigm with theories of some other evolutionary forces and processes. It is the task of paleobiologists to see whether such data can be found; and it is the task they have always taken very seriously, although they may arrive at sharply contrasting conclusions. This book is in fact devoted to their ongoing debate on this topic.

Every scientist is of course free to employ whichever methodological criteria best suit his or her taste. I myself have opted for pragmatic reductionism because it is an extension of Occam's Razor—the old philosophical principle that, while undertaking to describe and interpret the nature of the world, entities should not be multiplied beyond necessity—which has for centuries constituted the methodological foundation of modern science. It should by now be obvious that this option of mine must have tremendous implications for the position I take in the arguments on evolution that are the subject of the following chapters.

MACROEVOLUTION

6 What Is Macroevolution?

This book's goal is to present, discuss, and critically evaluate the significance of new ideas on evolution put forth in the 1970s and 1980s by my fellow paleobiologists. These ideas were developed in several books and dozens, if not hundreds, of academic articles. They also received a very wide, and generally sympathetic, hearing in popular media—from daily newspapers and television to illustrated weekly and popular science magazines. Certainly, the single most important corollary of all these writings is that the neodarwinian paradigm should be replaced with a more complete view of evolution, for although the genetical theory adequately explains microevolution, macroevolution is to be explained by a set of specific, macroevolutionary laws; one or more macroevolutionary theories must be developed, and hence, one of the main tenets of neodarwinism is invalid.

To analyze the meaning and validity of this claim, an understanding of the term "macroevolution" is obviously crucial. Wherever we deal in science with rival theories, the domain of their applicability must first be delimited in order to choose between them. To make a decision about competing historical explanations, the phenomena to be explained must first be identified. What are, then, the entities subject to processes described by the postulated macroevolutionary laws? What is macroevolution?

This term is, unfortunately, highly equivocal. Even within a single debate, discussants often use it variously. There are many explicit but sometimes sharply different definitions available—not only in articles by either advocates (e.g., Eldredge 1979, Gould 1983) or opponents (e.g., Charlesworth et al. 1982, Levinton 1983) of the macroevolutionary thinking, but also in textbooks (e.g., Raup and Stanley 1978, Futuyma 1979, Luria et al. 1981, Minkoff 1983). The situation is further confounded by history, because the concept of macroevolution has undergone a considerable change since the time Philiptschenko (1927) introduced and Goldschmidt (1940) made popular a sharp distinction between micro- and macroevolution.

Richard Goldschmidt fully accepted the neodarwinian postulate that natural selection is the main force responsible for evolutionary change within populations and species. He believed, however, that no genetical force envisaged by the neodarwinian paradigm can bring about a new species, let alone a new species that could be regarded as giving origin to a new genus, family, order, etc. In his view, some novel genetic mechanisms had to be invoked to explain the fact that such events must have repeatedly taken place in the history of the biosphere; for he of course believed in evolution as the mechanism of species origination. Goldschmidt indeed proposed that such evolutionary change beyond the limits of a species could only be produced by macromutations, or genotypic change with extraordinarily profound phenotypic effects. He thus distinguished between the phenomena and processes of microevolution, on the one hand, and the phenomena and processes of macroevolution, on the other. Microevolution entailed all evolutionary phenomena within populations and species, brought about by natural selection, genetic drift, and other forces acting in accord with the laws of neodarwinian population genetics. Macroevolution, in turn, included the appearance of new species, and hence also of new higher taxa, due to macromutations of various sorts. Goldschmidt's ideas provided the genetic basis for the theory of typostrophism, developed by the great and highly influential German paleontologist Otto Schindewolf (1950) who also treated natural selection as responsible only for microevolution, or the origin of geographic races of organism, but not for macroevolution.

A distinction between evolutionary processes within closely related groups of organisms (species?) and biological processes occurring on a grander scale was in fact nothing new. Long before Darwin, J. C. M. Reinecke (1818), a German student of ammonites, accepted evolution within individual groups of these fossil cephalopods and even claimed that organic evolution had not yet ceased; he nevertheless did not imagine an evolutionary transition between major groups of organisms. The paleontologist and biostratigrapher Friedrich August Quenstedt (1852) arranged ammonite fossils coming from successive strata into long, gradually and continuously changing series which he interpreted as the record of phyletic evolution driven by an intrinsic biological force and analogous to the ontogenetic transformations of individuals. But he never—not even a quarter of a century after Darwin—expanded this view to encompass series of similar genera or families. This tradition was further developed by another German student of ammonites, Wilhelm Waagen (1869), who coined the term "mutation" to designate series of slightly different variants of a single species derived from successive strata and thus representative of evolution. He, too, restricted the concept of evolution to small groups of closely related organisms and explicitly refrained from applying it to more dramatic differences between organisms.

All these paleontologists looked very closely at the fossil record and directly interpreted their empirical observations. They perceived continuous and gradual series of stratigraphically arranged fossils and explained them by evolution. Such series, however, could be constructed solely at the species level, not at higher taxonomic levels; hence, these paleontologists must have made a distinction between processes responsible for changes between and within species in time. This perspective on life's history was later taken up, and expressed in the form of a sharp dichotomy between micro- and macroevolution, by the geneticist Goldschmidt, though his observations concerned of course intraspecific genetic and phenotypic variation in living organisms.

In his *Tempo and Mode in Evolution,* George Gaylord Simpson (1944) employed Goldschmidt's terminological distinction between micro- and macroevolution, though he shifted the demarcation line by including speciation phenomena and processes to microevolution. He even introduced another term of this kind, "megaevolution," to designate evolutionary phenomena from the level of family upward. He employed these terms, however, solely for the sake of facilitating the communication among historical biologists concerned with description of various evolutionary phenomena. The main thrust of his entire argument was that micro-, macro-, and megaevolution are not qualitatively, or causally, different from one another. His point was that all historical biological phenomena—whether a pattern of temporal change in gene frequency in a fruitfly population, or a speciation event among frogs, or appearance of a new body plan (and hence a new higher taxon) in the biosphere—are brought about by the same set of evolutionary processes and hence explicable by the same set of evolutionary laws, namely those actually operating within populations and species.

The same point was made by Bernhard Rensch (1947), another eminent neodarwinian, although Rensch in speaking of the descriptive hierarchy of historical biological phenomena preferred to replace micro- and macroevolution with the terms "infra-" and "transspecific evolution," respectively. His conclusion was that all phenomena of transspecific evolution, such as large-scale evolutionary trends, distribution of evolutionary rates, and so on, arise by the infraspecific mechanisms of species formation and differentiation, driven by the genetic forces of evolution.

These were among the first and fullest expositions of the neodarwinian paradigm. Macroevolution was here explicitly distinguished for pragmatic purposes only, as a convenient term for the phenomena usually analyzed by paleontologists, in contrast to those studied by geneticists. When Simpson later realized that some biologists might be nevertheless misled by his application of Goldschmidt's terminological distinction between micro- and macroevolution, he suggested abandoning these terms altogether because they were likely to produce more confusion than clarity (Simpson 1953).

Simpson's opinion prevailed for two decades. The revival of emphasis on macroevolution in the 1970s and 1980s, however, not only brought the term back into the popular usage but also caused a shift in its meaning. It has begun to designate much more than merely the origin of evolutionary novelty, as meant by the older authors. For example, Niles Eldredge and Joel Cracraft (1980) define macroevolution as the pattern of any change in species composition of particular organic groups, both in space and time, and also as the process of differential species origination and extinction within higher taxa. On the other hand, they define microevolution as the pattern and process of change in gene content and frequency within populations. They thus exclude the phenomena and causation of appearance of new species and new body plans in evolution beyond the scope of distinction between micro- and macroevolution. In a report on interdisciplinary discussions on the role of ontogenetic development in macroevolution, Maderson et al. (1982) described macroevolution as encompassing both the patterns of differential origination and extinction of taxa, at any level of the taxonomic hierarchy, and the appearance of fundamentally new phenotypic features. Henryk Szarski (1986) explains the concept of macroevolution by providing two examples of macroevolutionary phenomena: the competition among species and higher taxa, and the origin of groups of organisms with new body plans, as for instance the reptiles and the birds. I myself used to consider the patterns of change in taxonomic diversity and ecologic structure of the entire biosphere as a typical macroevolutionary phenomenon (Hoffman 1986), whereas David Jablonski (1986a) and many other students of mass extinctions explicitly discuss the nature and impact of these phenomena on the biosphere as macroevolution.

Macroevolution thus means nowadays different things to different people. Instead of arbitrarily opting for one or another definition, I prefer to adopt a pluralistic approach and subsume them under one single heading. There is only one common denominator for all these different meanings: They all entail phenomena that can be described using species and higher taxa, rather than individual organisms or genotypes, as entities. Therefore, they all potentially call for evolutionary explanations that would go beyond the genetical forces envisaged by the neodarwinian paradigm.

Macroevolution is the pattern of supraspecific phenomena in space and time (Hecht and Hoffman 1986). Such phenomena include, for example, origins of fundamentally new body plans, changes in frequency distribution of particular phenotypic characters in various organic groups or even in the whole biosphere, distribution of rates and modes of evolution within and between organic groups, concordance versus discrepancy between evolutionary histories of major taxonomic groups, rates of species (or higher taxa) origination and extinction and their relationship, if any, to diversity in various groups, ecosystems, or even the biosphere. These phenomena are beyond the time scale of human life. Hence, they cannot be directly observed and experimented with but only derived—through pattern recognition—from the fossil record as a source of historical data on evolution.

A subset of these phenomena, those encompassing the grandest possible biological scales, the whole biosphere or at least a substantial realm of life, can be distinguished as megaevolution (Hecht and Hoffman 1986).

Whether such macro- and megaevolutionary patterns can be causally explained by microevolutionary processes, or whether they can only be adequately explained by reference to some specific macro- and megaevolutionary laws, describing the action of evolutionary forces complementary to, or superimposed upon, those envisaged by the genetical theory—this is the question at issue in the arguments on evolution I am now, at last, going to discuss.

Many paleobiologists are interested primarily in macroevolution. Some of them, perhaps the most vocal ones, maintain that macro-evolution is indeed causally independent at microevolution. They claim to have discovered macroevolutionary laws. This claim is the bone of contention.

7 Punctuated Equilibrium: A Multitude of Interpretations

7.1 The Slogan and Its Meanings

The modern history of macroevolution—the currently ascending wave of emphasis on macroevolutionary processes as causally distinct from those driven by the interplay between the genetic evolutionary forces and environment, and the consequent claims for inadequacy of the neodarwinian paradigm—began with the advent of punctuated equilibrium.

Fifteen years ago, Niles Eldredge and Stephen Jay Gould, two then young but now prominent American paleontologists, published an article (Eldredge and Gould 1972) that triggered an unusually hot debate, a debate that continues today. At issue is Eldredge and Gould's proposal that the view of evolution they regarded as the neodarwinian paleontological tradition and labelled "phyletic gradualism" is all wrong and should be replaced by another one, which they outlined in their paper and called "punctuated equilibrium." The implications of this proposal raise questions not only about the evolutionary inference from paleontological record but also about the evolutionary theory as such. The theory of punctuated equilibrium led Seven Stanley (1979) to claim that, "Macroevolution is decoupled from microevolution." And punctuated equilibrium prompted the titles like "Evolutionary theory under fire" (Lewin 1980) to appear even in such top scientific journals as *Science*.

The impact of punctuated equilibrium on evolutionary theory as a stimulus to consider and attempt to explain not only microevolutionary phenomena but also macroevolutionary ones, to be derived from the fossil record, is unquestionable. It has thus brought the problem of the relationship between neodarwinism and macroevolution back to the focus of modern evolutionary biology. The effects it has had on paleontology, however, go far beyond provoking a discussion on the tempo and mode of evolution as documented by the fossil record. The proposal of punctuated equilibrium sent a shock wave through evolutionary paleontology that still is clearly felt today. For it has shown that some established patterns of evolutionary interpretation of paleontological data are merely conventional stereotypes that can, and even should, be shaken off if we are ever to achieve a better understanding of what the fossil record really tells us about evolution; it has shown that much higher than traditionally accepted standards of pattern description and analysis are necessary if paleontologists are ever to be capable of learning about the tempo and mode of phenotypic evolution in the geological past; and it has also shown that evolutionary paleontology can, and even should, not only describe various extinct organisms and reconstruct the phylogeny of various organic groups but also examine questions that are of direct interest to modern evolutionary biologists. No doubt, punctuated equilibrium has thus enormously helped to elevate the status of paleontology as a science. It has increased the hopes, but also the requirements, associated with research in this field. But, somewhat ironically, although these truly great credits must not be denied to punctuated equilibrium, it fails, in my view, to meet the expectations of something really new and exciting it has itself stimulated.

During the long and often fervent debate on punctuated equilibrium, this concept has undergone a rather confusing evolution from its original statement to the most recent formulations. The term—like the term "macroevolution," with which it is associated—has come to mean very different things to different people, and even its authors have apparently employed it in quite different ways. The confusion is perhaps best illustrated by the contrasting opinions expressed by two prominent critics, Richard Dawkins and Olivier Rieppel. The former regards punctuated equilibrium as merely as minor wrinkle on the body of the neodarwinian paradigm, a little variation situated firmly within the patterns of evolutionary thought established by Darwin and the Modern Synthesis of the 1940s, which has only been blown out of proportion by irresponsible journalists (Dawkins 1986). The latter, by contrast, considers punctuated equilibrium to be a revolt against the principle of continuity—one of the most fundamental philosophical principles underlying not only the neodarwinian paradigm but also the entire tradition of modern natural sciences, going back at least to the great seventeenth-century philosophers Descartes and Leibniz (Rieppel 1987). When two bright and competent people describe a single concept in terms so sharply different, the concept itself must be faulty in that it is imprecise and hence equivocal. To clarify this semantic tangle may be a rather pedantic task, but this job must first be accomplished before the validity and significance of punctuated equilibrium can be discussed and evaluated.

In their original article, Eldredge and Gould proposed that the microevolutionary processes envisaged by the neodarwinian paradigm should, and indeed generally do, result in a macroevolutionary phenomenon unexpected by evolutionary biologists and paleontologists. This phenomenon is a distinctly bimodal frequency distribution of genotypic, and consequently also phenotypic, rates of species evolution along a phyletic lineage, within a succession of ancestors and descendants. Eldredge and Gould rejected the idea—which they attributed to phyletic gradualism and regarded as orthodox within the neodarwinian paradigm—that the evolution of species generally is a slow and gradual change leading over very long time intervals in the same adaptive direction; for example, toward increased shell globosity in a group of ammonites, or toward decreased adult body size of elephants on a small island. In their view, phyletic gradualism maintains that evolution entails primarily a continual adaptation of each species to unidirectional change in at least one ecological parameter of its effective environment, for instance, water temperature in a marine planktic ecosystem or top predator swiftness on a savannah. This evolution, furthermore, is expected to proceed at a slow and constant rate. New species and even new genera are expected to always originate by such transformations of old species. The predicted macroevolutionary pattern is therefore a linear trend in phenotypic characters on the geological time scale.

According to Eldredge and Gould, such evolution is explained by neodarwinian paleontologists, from George Gaylord Simpson on, by orthoselection, that is, by the kind of evolutionary force of natural selection that operates directionally and pushes each population toward acquisition of adaptations necessary to adequately meet the challenge of continually changing environment. Phyletic gradualism, then, simply substitutes the neodarwinian concept of orthoselection for the outdated and somewhat mystical concept of orthogenesis, or unidirectional evolution of organic groups caused by their intrinsic drive toward some specific adaptive goals, which was particularly fashionable among paleontologists in the late nineteenth and the early twentieth century. But it is the fundamental premise of this paleontological tradition that the macroevolutionary pattern of phyletic gradualism to be explained by either orthogenesis, or orthoselection is indeed pervasive in nature.

Were it so, however, the neodarwinian paradigm would be in real trouble. Eldredge and Gould wrote that, on the one hand, the concept of orthoselection is largely untestable and hence a poor explanation for the presumed evolutionary pattern. On the other hand, phyletic gradualism contradicts the neodarwinian paradigm that actually predicts that the evolutionary history of particular phyletic lineages should mainly consist of very long time intervals of stasis—when essentially no evolutionary change takes place—interrupted from time to time by rapid speciation events. Speciation occurs because small, isolated populations, at the periphery of the geographic range of a species, are subject to extreme selective pressures in atypical environmental settings they encounter, and therefore natural selection leads to their evolution toward adaptations different from those advantageous for the main populations. The large mainstream populations do not evolve in the meantime because their environment remains essentially the same as it used to be, and also because their sheer size confers upon them a certain evolutionary inertia, or resistance to change, induced by some intrinsic, biological, chiefly genetic, homeostatic mechanisms. As a result, the marginal populations become genetically, and also phenotypically, separated from the formerly conspecific populations, with which they previously constituted a single species. This is the allopatric mode of speciation, advocated mainly by the great neodarwinian Ernst Mayr (1963), who used to consider it to be the prime, if not the only, way of cladogenesis, or the origination of new phyletic lineages.

According to Eldredge and Gould, the neodarwinian paradigm of evolution thus predicts that the microevolutionary forces result in the macroevolutionary pattern of punctuated equilibrium, instead of the one of phyletic gradualism. It implies that the norm in evolutionary history of any species is genetic and phenotypic homeostasis, which can break down solely by allopatric speciation in small, marginal populations at the periphery of the species's range. [In a later article,

Gould and Eldredge (1977) also accepted a possibility, and even plausibility, of sympatric speciation—without actual occurrence of a geographic isolation between the population evolving in a new and unique adaptive direction and the rest of the species—but this problem is without much significance for either the validity or the implications of punctuated equilibrium.] In this view, then, it follows from the neodarwinian paradigm, or at least from its Mayrian version, that almost all evolutionary change should be concentrated in speciation events, whereas the change taking place within particular species, between successive speciation events, should be negligible and virtually nonexistent.

The frequency distribution of evolutionary rates should therefore be distinctly bimodal—very high rates at speciation and zero rates at other times—but such a macroevolutionary pattern of bimodality in rates runs counter to the entire paleontological tradition of evolutionary inference from the fossil record. As put by Eldredge and Gould, however, this is indeed the pattern of evolutionary rates revealed by paleontological data; paleontologists document almost exclusively very long-ranging species separated from each other by large phenotypic gaps reflecting discontinuities produced during speciation. The fossil record of evolution shows mainly stasis, interrupted occasionally by very rapid change—hence, the slogan "punctuated equilibrium."

Clearly, punctuated equilibrium was originally formulated in opposition to phyletic gradualism, the latter being understood as unidirectional evolution at a slow and constant rate, caused by orthoselection affecting the entire species. At the same time, however, punctuated equilibrium portrayed speciation as a necessary condition for any significant evolutionary change in organisms. Consequently, it could be, and indeed used to be, variously interpreted.

The situation has been further confounded by the multifarious corrections, reformulations, and reinterpretations made later by Eldredge, Gould, and their main allies, Steven Stanley and Elisabeth Vrba. To the neodarwinian Mayr, speciation—even a very rapid one—is always a populational process that must take a number of generations and hence, time. This was also the position Eldredge and Gould explicitly took in their original article. In theory at least, the pattern of punctuated equilibrium should therefore boil down to an alternation of periods of stasis and acceleration of the rate of evolution at speciation, but it should not indicate a real disruption of continuity of the process. It is only in the fossil record that the appearance of discontinuity between the ancestral and the descendant species may arise, because the very rapid evolution leading to speciation is expected to take place in populations that are so small they are unlikely to ever be preserved in the record.

In a later article, however, Gould and Eldredge (1977) implied that the boundary between ancestral and descendant species was always discontinuous and hence very real. They referred punctuated equilibrium to the cultural tradition that puts emphasis on a fundamental difference between minor quantitative changes and rapid qualitative leaps from one steady state to another. They explicitly invoked Karl Marx and Friedrich Engels in this context, as the most prominent representatives of this tradition of rejecting the principle of continuity. They thus clearly expanded the scope of punctuated equilibrium to bear more on the true process of transition from one species to another, and not only to reflect those qualities of the fossil record that determine the differential likelihoods of preservation of very large and very small populations.

In a similar vein, although without ever referring to Marxism, Stanley (1975) used the term "rectangular evolution" as a synonym for punctuated equilibrium. He meant that the phenotypic evolutionary change at speciation should take place essentially instantaneously, causing a jump from one phenotype to another. Both Stanley and Gould later developed this view of evolution much further, presenting speciation events as very rapid flips between fairly stable equilibria normally maintained by the intrinsic stability of phenotypic structures and their consequent resistance to any change. Minor adjustments and modifications are possible in the vicinity of each equilibrium—and this is what natural selection normally achieves—but transitions between these domains of stability are very difficult and occur in a different way, by breaking down the homeostatic mechanisms of ontogenetic development and hence disrupting the continuity of evolutionary process (e.g., Gould 1980, 1982a). As Gould (1984) put it in the form of analogy to Marxism, "Heat water, and eventually it boils. Bend a beam, and eventually it breaks. Opress the workers more and more, and eventually they revolt." This is the way a real change is expected to occur, both in the social and in the organic world.

Punctuated equilibrium thus leads to a new viewpoint on evolution. It emphasizes the active resistance of biological structures to evolutionary change. It rejects the notion of gradual, continuous evolution, because each species has its own domain of phenotypic stability within which minor changes are likely to be effected by natural selection but which cannot be easily abandoned in the process of evolution. Such domains of phenotypic stability are delimited by the coherence and integrity of structures, brought about in the course of ontogenetic development, and by the constraints they impose on evolutionary change that could, in principle, be effected by the genetic evolutionary forces (e.g., Gould 1980, 1982b). According to this view, there are very severe topological limitations to the ways an embryo can undergo transformations during ontogeny; a shape or structure can be changed into some, but not all, other conceivable forms. Consequently, there are limits to the variety of adult phenotypes that can develop. Natural selection's potential for evolution toward adaptations is therefore strongly constrained by development. It is largely these developmental constraints that determine the pattern of phenotypic evolution, whereas natural selection and the other genetic forces appear as merely subordinate factors. Evolution thus depends primarily upon the resistance of developmental programs of species to change. Such a change can occasionally be effected by macromutation, and this is why the name of Richard Goldschmidt was invoked in this context by Gould (1980), who even coined the term "Goldschmidt break" for the causal discontinuity he envisaged between speciation and microevolutionary processes within species.

But if substantial phenotypic change can be only with much difficulty accomplished in evolution, then stasis, or phenotypic stability of species, should be a pervasive feature of the pattern of evolution. Organic species each acquire therefore a clearcut identity. They are no longer arbitrarily isolated segments of evolving phyletic lineages, but they originate very rapidly by speciation events and then persist essentially unchanged until they go extinct; and they can give rise to new species at any stage of such evolutionary history. Every species is thus easily recognizable as a separate element of the pattern of evolution. Every species is an independent player in the evolutionary game, and it is the fate of species rather than the history of phenotypic transformations that evolution is in fact all about.

According to both Eldredge (1985) and Gould (1982c, 1985), this is in fact the most important corollary of punctuated equilibrium. On the one hand, it raises the question, how can species stasis be maintained in spite of the incessant environmental change? On the other, it points to a potential problem with neodarwinian explanations for some macroevolutionary patterns. For if species, once established, no longer undergo significant evolutionary change, how should we explain long-term evolutionary trends, or sequences of fossil populations with morphologies shifting consistently in the same direction— for example, toward a reduction of the number of toes in the horse foot—over millions and millions of years?

One possible answer is that each of such sequences actually consists of a discontinuous succession of species, with each species having constant morphology throughout its temporal range and with the appearance of trend arising solely from the preconceived notion of continuity linking them all together into a single evolving, biological entity. Since organic species, however, are portrayed within the framework of punctuated equilibrium as individual entities, they can in principle be subject to selection *of,* perhaps due to selection *for* some their properties. Punctuated equilibrium may therefore suggest the macroevolutionary theory of species selection as the explanation for evolutionary trends. And this theory can be further expanded into a new perspective on evolution—the hierarchical paradigm, which envisages a whole hierarchy of the levels of evolution, and also selection, from individual organisms at the bottom to populations and species and up to monophyletic clades, or higher taxa including all the species descendant from a single ancestral one (Eldredge 1985, Salthe 1985).

Clearly, punctuated equilibrium has become a general heading for a complex nexus of interrelated but distinct ideas on evolution, rather than a particular model, hypothesis, or theory. Depending on the emphasis put on one or another component of this nexus, it is prone to a multitude of interpretations. A sort of typology of these interpretations is in order, for it will help to assess the place and the role of punctuated equilibrium in modern evolutionary theory. This may sound pedantic but I distinguish three main versions of punctuated equilibrium, and also a couple of variants of these versions. Thus, I discuss in turn the weak, the mildly and radically strong, and the skeptically and enthusiastically moderate interpretations of punctuated equilibrium.

7.2   The Weak Version

The original statement of punctuated equilibrium could be simply understood as rejection of phyletic gradualism, that is, as negation of universality, or even merely commonness, of unidirectional evolution of species going on with approximately constant pace and gradually transforming the entire ancestral species into its descendant one. This interpretation is what Max Hecht and I have termed the *weak* version of punctuated equilibrium (Hecht and Hoffman 1986). It means that, generally, the evolution of species does not proceed for a long time in the same direction and at the same rate, but that it varies both in direction and in rate.

That this weak version is not an intellectual straw man, artificially extracted from the original article by Eldredge and Gould (1972) but never meant seriously as a theoretical proposition, is well evidenced by the recent articles by Niles Eldredge (1984) and Elisabeth Vrba (1985a). Vrba explicitly writes that the concept of punctuated equilibrium tells only that the rate of species evolution should fluctuate along particular phyletic lineage, and such patterns should be found to prevail in the fossil record; and she contrasts punctuated equilibrium with the views of phyletic gradualists whom she describes as accepting a variation in evolutionary rates between lineages but contending that the rate of evolutionary change should be constant within particular species as well as along their phylogenetic sequence. Eldredge, in turn, describes the idea of punctuated equilibrium as stating only that some speciation events entail also a detectable phenotypic change, but fully allowing for significant evolutionary change occurring also between speciation events, that is, within particular species.

This weak version of punctuated equilibrium is, of course, nothing new in evolutionary theory. It is entirely trivial within the neodarwinian paradigm of evolution, because neodarwinians never did stick to phyletic gradualism as defined by Eldredge, Gould, and Vrba. This assertion is abundantly documented by the classic works of George Gaylord Simpson (1944, 1953) who painstakingly demonstrated that evolutionary rates do vary; and the great geneticist Hermann J. Muller (1949) established this view as a cornerstone neodarwinian synthesis in his summary of the interdisciplinary symposium in Princeton, which is widely held to have been a sort of official proclamation of the neodarwinian paradigm. This should not be at all surprising because Simpson's main effort, and also his great achievement, was to combat phyletic gradualism. He sought to show that the fossil record of evolution is fully compatible with the neodarwinian explanation by the interplay of natural selection and genetic drift with environment, whereas the advocates of orthogenesis indeed presented the fossil record in phyletic gradualist terms and claimed it to refute the theory of natural selection. As quite rightly noted by Eldredge and Gould (1972), the neodarwinian paradigm does not predict the linear pattern of evolution; but, then, their assertion that the pattern indeed is not linear appears to be anything but revolutionary. In fact, even a critical reading of Charles Darwin's own writings proves beyond any reasonable doubt that the picture of evolution as a stately unfolding, unidirectional, irreversible change at a constant rate, does not follow from his ideas about how evolution proceeds (Penny 1983, Rhodes 1983).

One might argue that regardless of what Darwin or Simpson did or did not actually write on the tempo and mode of species evolution, the universality of phyletic gradualism was postulated by the version of neodarwinism that dominated in American evolutionary biology, or at least paleontology, in the 1950s and 1960s, when Eldredge and Gould were students. Punctuated equilibrium originated in that sociological and intellectual context, and it might constitute a rebellion against its authors' neodarwinian teachers and peers.

Eldredge and Gould indeed seem to strongly support such interpretation, for in their original article they devoted much attention to the historical context of punctuated equilibrium. In outlining this context, however, they mainly focused on the classic nineteenth-century textbooks: English translation of the monumental treatise by the German paleontologist Karl von Zittel (1876–1893) and successive reeditions of *Elementary Paleontology* by Woods (1893). It is of course quite natural that the picture of evolution conveyed by these textbooks is very different from the neodarwinian paradigm, whose inception they precede by many decades. They were written when the vast majority of biologists and paleontologists rejected Darwinism and resorted to a variety of orthogenetic and Lamarckian concepts, in part because they could not reconcile the evidence of phyletic gradualism they saw in the fossil record—whether rightly or not—with natural selection as the prime evolutionary force. To rebel in 1972 against the antidarwinian tradition personified by Zittel and Woods should not strike anyone as particularly bold and innovative.

In fact, even the classic American textbooks of the middle of the twentieth century cited by Eldredge and Gould (1972) did not advocate the extreme phyletic gradualism rejected by the weak version of punctuated equilibrium. For example, Moore, Lalicker, and Fischer (1952) wrote in their *Invertebrate Fossils*—the standard academic textbook of paleontology in the fifties and sixties, widely quoted by

Eldredge and Gould—that periods of slow and gradual evolution alternate in the history of individual lineages with bursts of explosive evolutionary change leading to marked transformations of both form and function. They noted some fossil groups change so slowly with time that they may seem not to change at all, while others undergo striking evolutionary modifications with geological abruptness. They also considered very rapid allopatric speciation in small, isolated populations as the main mode of species origination. Thus, the weak version of punctuated equilibrium appears as a mere restatement of what the neodarwinian has to say, and indeed was saying, on the rates of species evolution.

7.3   The Strong Version

The 1972 article by Eldredge and Gould, however, can also be interpreted as the proposition that gradual phenotypic change is virtually absent from the evolution of species, and that periods of complete evolutionary stasis of species are interrupted solely by speciation events that bring about a significant phenotypic change. Gradual change is here understood as entailing a continuous shift, through a series of intermediates, of the range of intraspecific variation in the considered phenotypic features. This shift does not necessarily have to be unidirectional or to proceed at a very slow and constant rate, but it must be detectable on the geological time scale and it must also eventually exceed the limits set to a single species by the usual taxonomic standards. Speciation events, in turn, are here understood as either the splitting of a phyletic lineage—be it by allopatric or by sympatric speciation—or the quantum evolution of Simpson (1944), who envisaged a mechanism that so greatly accelerated the rate of evolution that a population can rapidly acquire some phenotypic characteristics significantly different from those present in its ancestors; such a population can then be called, by convention and comparison to the usual taxonomic standards, a new species. The message of Eldredge and Gould can then be read as follows: Patterns of gradual evolution are nonexistent, or at best negligible; patterns of punctuated evolution, with nonbridgeable gaps between successive species, are overwhelmingly dominant.

This interpretation constitutes the *strong* version of punctuated equilibrium (Hecht and Hoffman 1986). Several articles by Gould

and Eldredge (1977; Gould 1980, 1982a, 1982c) and especially Stanley (1979, 1982a, 1982b) clearly demonstrate that this strong version
does not distort or exaggerate the meaning that was indeed attributed
to punctuated equilibrium by its proponents.

The strong version of punctuated equilibrium has been repeatedly
tested by paleontologists over the last dozen years or so. Unfortunately, such empirical tests in the fossil record can almost never be
decisive because, generally, the paleontological data do not allow for
unequivocal identification of gradual and punctuated modes of evolution of species. In order to demonstrate beyond reasonable doubt the
occurrence of gradual evolution in a geological setting, it is not
enough to find a gradualistic pattern in the fossil record, that is, to
show that an organic group is represented in a geological section by a
continuous sequence of fossil populations that exhibit significant phenotypic changes over a certain time interval. The process of gradual
species evolution is potentially a good causal explanation for such a
pattern. This gradual sequence, however, must also be shown to
encompass a true biological species rather than an arbitrarily isolated
group of individual organisms making part of a much larger population. It must also recur over an area that is sufficiently large and
ecologically heterogeneous to permit ruling out an alternative explanation for the pattern of gradual phenotypic change through geological time. It must rule out migration of a set of conspecific populations, or closely related species, representative of so-called clinal
variation, that is, exhibiting a geographic gradient in a given phenotypic character, and tracking their preferred habitats which wander in
response to a climatic or other environmental change.

This is a very difficult, often impossible, task because it requires a
very complete and detailed biological background knowledge to help
identify, by analogy to living organisms, the limits of intraspecific
variation; and it also requires a very precise time correlation of biological events taking place in a variety of areas. Yet this kind of
information is only rarely available for case studies in the fossil record. It is nevertheless feasible, at least under ideal circumstances, to
corroborate gradual phyletic evolution as the explanation for a paleontological pattern.

It is also possible, although again very difficult, to refute a gradualistic interpretation of paleontological data. To this end, however,
it is not enough to document a punctuated pattern of fossil populations. Provided that the geological record is episodic and contains

significant time gaps, such patterns may in fact be compatible with the process of gradual species evolution; for, it is entirely conceivable, and even likely, that the intermediate populations that are absent from a punctuated pattern were actually present in the considered area but are not preserved in the fossil record because of the incidental vagaries of its geological history. But if a temporal sequence of fossil populations exhibits considerable morphological gaps between samples of successive populations—gaps equivalent to phenotypic differences larger than the change accomplished along the continuous segments of the sequence—and if the geological section can be demonstrated to be continuous in time—to lack no significant time intervals—the gradualistic interpretations cannot hold and must be rejected. The problem is only how to show with reasonable certainty that the considered geological section is complete in terms of the time it represents. This can be done only if a pattern caused by gradual species evolution is discovered in the same geological section. Such coincidence, however, cannot be expected to occur commonly.

There is an important methodological asymmetry inherent in testing for gradual versus punctuated evolution of species. Gradual evolution can be both refuted and corroborated, though not ultimately proven. This is not the case with punctuated evolution. It cannot be, even in principle, corroborated as the explanation for a punctuated pattern discovered in the fossil record. Even granting the temporal completeness of a geological section comprising such a pattern, there is no way to provide a strong evidence, let alone to prove, that two successive populations separated by a considerable morphological gap indeed represent a single phyletic lineage; that they are linked by a genealogical ancestor–descendant relationship. There is also no way to rule out a possibility that a punctuation, or the first appearance of a population with radically new morphology, represents immigration of a species that originated earlier and evolved in a gradual manner but in another geographical area.

On the other hand, the strong version's claim that significant phenotypic change is always associated in evolution with speciation can hardly ever be addressed in the fossil record. I am aware of only a couple of really good cases of direct paleontological documentation of branching of a phyletic lineage, which provide indeed a compelling evidence for cladogenetic, speciation event (Grabert 1959, Prothero and Lazarus 1980, Lazarus 1986, Sorhannus et al. 1988); in addition they all deal with marine microfossils that have a uniquely high-quality fossil record but rather poorly known biology. But apart from these very unusual cases, speciation is identified in paleontology with profound morphological change between successive fossil populations. The association of such "speciation" with phenotypic change becomes then inevitable, although it may well be apparent, due to an arbitrary convention, rather than real.

It might be easier to refute punctuated equilibrium as the explanation for a paleontological pattern, but no matter how long and insensibly graded is the chain of successive populations in a geological section, one may always argue that it is better explained by a large number of periods of stasis and punctuations than by continuous evolution. The problem is the same as with any series of points on a graph, which can represent a curve as well as a broken line. There is no way to resolve such a dilemma objectively until a criterion is given that determines the minimum duration of a period of stasis or the minimum gap to be interpreted as punctuation. Such criteria, however, can only be established by arbitrary conventions, not by inference from any biological data or theory.

The empirical difficulties with documenting gradual phenotypic evolution of species in the fossil record are the reason the vast majority of the classic gradualistic examples described by paleontologists in the nineteenth and early twentieth century are now regarded as inadequately corroborated. Simply, the standards valid in those times are too low by comparison to what is presently required from such case studies. Nevertheless, and contrary to repeated assertions by the advocates of punctuated equilibrium, the modern paleontological literature, written after the standards were raised by the advent of punctuated equilibrium as an alternative to phyletic gradualism, does contain a considerable number of convincing instances of gradual phenotypic evolution. A sizable bibliography of such examples is provided by Hoffman (1982), Levinton (1983), Gingerich (1985), and Hecht and Hoffman (1986), and further examples continue to appear (e.g., Chaline and Laurin 1986, Chaline 1987, Sheldon 1987). They encompass a very wide variety of organic groups inhabiting a broad spectrum of natural environments—from pelagic plankton of the open ocean (foraminifers, radiolarians) through marine benthos to freshwater fish and land mammals—and concern changes in size as well as in shape of the organism. They generally meet the conditions of adequate sampling, appropriate biometric and statistical analysis, and sufficiently broad geographic and eco-logic coverage to rule out alternative explanations.

The main problem that often remains and may undermine the validity of these gradualistic cases is whether the extent of the observed phenotypic change is sufficiently large to warrant the claim that it is evolutionary in nature. This problem can only be overcome by application of analogy to closely related living organisms, where the range of intraspecific phenotypic variation is well known. For example, Paul Koch (1986) studied the geographic variation in modern land mammals and concluded on this basis that the gradual phenotypic changes observed by Philip Gingerich (1976; see also Bookstein et al. 1978) in Eocene mammals are much larger in extent than intraspecific variation; hence, they are likely to represent evolution rather than migration of conspecific populations or ecophenotypic response to a shift in some environmental parameters. Such support by analogy is even more straightforward in the case of the Plio-Pleistocene rodent *Mimomys* and other small land mammals studied in Europe by Chaline (1987) or the Miocene stickleback fish *Gasterosteus* from North America investigated by Bell and Haglund (1982; see also Bell et al. 1985). In both these cases, the authors themselves have an extensive knowledge of biology of the living relatives of the fossil organisms they consider.

This approach is much less productive, however, when applied to planktic foraminifers or radiolarians, simply because their biology is at present too poorly known to allow for assessment of their potential for ecophenotypic, as contrasted to evolutionary, change. Unknown is even whether they reproduce sexually or asexually, and the meaning of species in these groups is entirely obscure. Obviously, analogy to living relatives cannot be reasonably applied to entirely extinct groups, like the Jurassic ammonite *Kosmoceras* studied first by Roland Brinkmann (1929) and reanalyzed recently by Raup and Crick (1981). After a thorough analysis of biometric data on hundreds of specimens from a very well sampled section, Raup and Crick conclude that the gradual mode of phenotypic change of *Kosmoceras* in the investigated time interval is undisputable. But there exists no criterion to determine whether this is a case of true evolution, or merely minor oscillations within the limits of species stasis. One may thus speak of a whole spectrum of examples of gradual species evolution—from well documented to more tentative ones.

By contrast, all the known instances of allegedly punctuated evolution (e.g., Williamson 1981, Kelley 1983, Cronin 1985, Cheetham 1986) are, and must be, only tentative. Moreover, they are essentially confined to shallow-water habitats. Yet this is exactly the kind of depositional environment where the fossil record is bound to be particularly episodic and to comprise a significant number of time gaps. Every minor oscillation of the sealevel or the shoreline may cause in such environments erosion and thus introduce a gap to the record. Such time gaps, in turn, naturally translate into morphological gaps between successive samples from a continuously and gradually evolving lineage. It may not be pure chance that the less complete the stratigraphic record in certain sedimentary environments, the more cases of allegedly punctuated evolution of species are reported. Michael McKinney (1985) has shown a remarkable statistical correlation to this effect.

The notorious incompleteness of the fossil record may indeed largely account for the apparent morphological gaps between ancestral and presumably descendant species, as Charles Darwin so ingeniously claimed long ago, and as his successors invariably maintained ever since. A punctuated paleontological pattern does not have to reflect evolutionary punctuations by speciation events. On biological grounds, there seems to be no relationship between the rapidity and extent of evolutionary change and speciation. The available genetic information proves beyond any reasonable doubt the absence of any unequivocal, one-to-one relationship between phenotypic, and also genetic, change and speciation; speciation may be associated with very large as well as very little change (see e.g., Ayala 1975, Douglas and Avise 1982).

In principle, an overwhelming dominance of the punctuated mode of species evolution, as postulated by the strong version of punctuated equilibrium, might also be inferred from bimodality of the frequency distribution of phenotypic rates of evolution—that is, had such bimodality been actually observed in the record. As shown by Philip Gingerich (1983), however, there is a continuous spectrum of evolutionary rates, from genuine stasis to extremely rapid evolution. This observation is further supported by very interesting, though not at all surprising, paleontological evidence for patterns of phenotypic change that are intermediate between the gradual and the punctuated ones. In several cases, the conditions are met to ensure that an evolutionary explanation for such patterns seems most appropriate. Granting such interpretation, in some fossil lineages the rate of phenotypic evolution underwent a considerable acceleration without any detectable relation to their splitting, or speciation (e.g., Cisne et al. 1982, Malmgren et al. 1983). This phenomenon has been called punctuated gradualism. In other lineages, speciation and the associated phenotypic change occurred relatively rapidly, but certainly gradually even on the geological timescale (Lazarus 1986, Sorhannus et al. 1988).

This brief discussion of the empirical paleontological data relevant to evaluation of phyletic gradualism and punctuated equilibrium does not lead to refutation of the proposition that punctuated phenotypic evolution of species does indeed occur in nature. All the critics of punctuated equilibrium agree that phenotypic punctuations do happen. No doubt, both the gradual and the punctuated modes of evolution certainly occur, quite likely even in the same lineage, at different stages of its evolutionary history. But this discussion shows that the strong version of punctuated equilibrium—the claim that gradual phenotypic evolution of species occurs at best exceptionally, orders of magnitude less commonly than punctuated evolution—is blatantly false. There is no empirical data to support it, and a considerable body of data to contradict it.

7.4 Having Accepted the Strong Version . . .

The advocates of punctuated equilibrium, however, took the strong version for granted and, in order to explain the postulated macroevolutionary phenomenon—the allegedly strict bimodality of evolutionary rates (very rapid speciation and stasis)—proposed a fundamental biological difference between the processes of microevolution and speciation. They invoked the theories of speciation put forth by Ernst Mayr, Hampton Carson, and Sewall Wright, three very prominent neodarwinians, to substantiate their own claim for possibility, and even necessity, of speciation being always very rapid, almost instantaneous, due to genetic revolution, which they often interpreted in a very Goldschmidtian way—as the occurrence of macromutations, or point mutations with very large phenotypic effects, leading to immediate origination of new high-rank taxa. Stanley's (1982c) interpretation of the origin of the class Gastropoda by a single mutation, which caused torsion of the visceral hump in a single monoplacophoran female, is a good example. In this view, then, speciation is a macroevolutionary phenomenon caused by a macroevolutionary process, qualitatively different from and irreducible to the processes of microevolutionary change. This is why Gould (1980) wrote of the neodarwinian paradigm as being "effectively dead."

Two different variants of this perspective on speciation by macromutations are conceivable. According to one variant, or the *mildly strong* version of punctuated equilibrium, speciation is caused by a separate class of evolutionary processes, but the principle of continuity is nevertheless observed in that the process must concern populations and conform to the laws of genetics. The other variant, or the *radically strong* version of punctuated equilibrium, maintains that speciation introduces a fundamental discontinuity to the evolutionary process and thus violates the principle of continuity. The two variants substantially differ from each other in their status, because the former refers, correctly or not, to some empirical and theoretical biological knowledge, whereas the latter reflects a metaphysical option that cannot be corroborated or refuted.

The mildly strong version implies that, in spite of their proverbial crudeness and incompleteness, historical biological data from the fossil record can force evolutionary biologists to undertake a far-reaching revision of their views on microevolution, macromutations, and speciation. This implication must have, and indeed has, provoked a rather strong reaction. For the pattern of punctuated phenotypic evolution would perfectly fit—even if it were indeed so common in the history of the biosphere as the advocates of punctuated equilibrium expect—a very wide variety of quantitative models of microevolutionary processes that could bring about a relatively rapid phenotypic shift from one evolutionary steady state to another (e.g., Kirkpatrick 1981, Petry 1982, Newman et al. 1985, Lande 1986). Without a capability of direct observation of, and experimentation with, the ongoing process, it is impossible to determine whether the process is adequately represented by one or another of these microevolutionary models. Yet the rate of change that can be accomplished by these ordinary neodarwinian processes—a combination of natural selection, genetic drift, and other forces—certainly exceeds the rates actually indicated by even the sharpest punctuations observed in the fossil record; given, of course, the liberal interpretation that these punctuations are indeed evolutionary events. For example, the most famous case of punctuated evolution has been described by Peter Williamson (1981) from a number of Plio-Pleistocene freshwater mollusks from

Lake Turkana in East Africa; these punctuations are now being reinterpreted by Williamson (1985) as speciation within perhaps 50,000 generations. No matter if the morphological change observed in these animals in the geological section at Lake Turkana can actually be regarded as true speciation—which I doubt, if only because there is no faintest evidence for occurrence of lineage splitting, while time relationships among various morphotypes of Turkana mollusks are ambiguous—50,000 generations is long time by any biological standard. It is orders of magnitude more than predicted by some microevolutionary models of speciation. Thus, the punctuated paleontological pattern, even when interpreted as a record of evolution, does not call for evolutionary explanation by a special class of evolutionary processes involving macromutations.

No wonder all the evolutionary theorists, whose concepts of speciation were invoked by Gould and Stanley to support their strong version of punctuated equilibrium, reject outright the perspective on speciation as decoupled from microevolutionary processes (Carson 1982, Mayr 1982, Wright 1982). For even if speciation were indeed always due to genetic revolution, or disruption of the genetic and developmental homeostatic mechanisms responsible for keeping the modal phenotype of a population in a kind of evolutionary steady state, it would nonetheless remain a gradual process of population change. Yet many evolutionary biologists also disagree with the assumption of universality, or even considerable commonness, of genetic revolution as prerequisite to speciation. The wide variety of genetic and ecologic models of speciation coexisting side by side on the marketplace of biological ideas indicates that speciation may sometimes occur rapidly, and sometimes it may not; it may sometimes involve a dramatic reorganization of the ancestral genome, and sometimes it may not. There seems to be nothing in the processes of speciation that would necessarily demand macromutations.

Macromutations most certainly occur in nature. In every population of organisms, individuals appear occasionally that quite dramatically deviate from the phenotypic norm. The fruitfly *Drosophila* with additional pair of legs replacing the sensory limbs growing normally on the head of the animal is a good example. This abnormality is known to be caused by a single point mutation in DNA, as demonstated long ago by the pattern of its heritability and confirmed by molecular studies. Such monstrous phenotypes are indeed quite often caused by point mutations, which are therefore macromutations.

These macromutations, however, are most likely to be adaptively disadvantageous, or even lethal as noted by Sewall Wright (1982). Wright examined thousands and thousands of guinea pigs and discovered dozens of macromutations, but not even one of them had any chance of survival and reproduction under normal ecological conditions. This is quite understandable. For, if natural populations usually are relatively close to their adaptive optimum, the larger the phenotypic effect of a mutation, the smaller is the likelihood it will decrease this distance. This rule is particularly true for phenotypic traits that are part of well-integrated functional complexes—like the vertebrate eye, for example—where every change is more likely to cause damage than improvement. To paraphrase the famous saying by Richard Goldschmidt, macromutations result in hopeless rather than hopeful monsters; though there is of course no logical way to disprove their occasional hopefulness.

In turn, macromutations known to lead to viable, or even adaptively advantageous, phenotypes generally do not result in speciation. In the salamanders *Ambystoma,* for example, individuals sometimes appear with considerably enlarged head and especially jaws and with cannibalistic behavior (Collins and Cheek 1983). This phenotype also is due to a single mutation, and it is capable of gaining dominance in a population under a certain atypical set of ecological circumstances. When the conditions return to the norm, however, this cannibalistic phenotype loses the edge and gives way to the more ordinary phenotype. Thus, no speciation takes place in spite of the occurrence of a macromutation. This is not an unusual phenomenon, as demonstrated, for instance, by John Turner's (1981) study of the mechanism of origination of mimicry in the butterfly *Heliconius* or by Allan Larson's (1983) genetic analysis of the origins of innovative phenotypic features in plethodontid salamanders. The process often seems to begin with a single macromutation followed by a cascade of minor changes.

Thus, the concept of macromutations as a distinct class of genetic events constituting the main mechanism of speciation appears today implausible, to say the least. Given this conclusion, however, and given also the absence of evidence of the overwhelming dominance of the punctuated mode of evolution, the mildly strong version of punctuated equilibrium is indefensible.

In fact, no one seems to be willing to advocate it any longer. In a recent article, Gould (1985) explicitly rejects this interpretation of his theory. He writes that what punctuated equilibrium is about is neither the distribution of evolutionary rates, nor the mechanism of speciation, nor the problem of macromutations and their role; and he refers the origination of new species to microevolution. He nevertheless still portrays punctuated equilibrium as embodiment of a general philosophy of change through time, the one he earlier (Gould 1984) described as "punctuational change to emphasize both the stability of systems and the concentration of change in short episodes that break old equilibria and quickly re-establish new ones," and which he then related to Marxism and all other ideologies stressing discontinuity, giving more weight to objects and less to continuous flux that links them all together in a causal chain. Still earlier, Gould (1980) compared the evolution of species to the notion of a polyhedron, which only can either slide on one of its faces (equivalent to stasis), or topple over to accomplish an evolutionary change literally by saltation; and he explicitly used the oxymoron "discontinuous evolution" in order to describe the latter process. Gould thus supports what I am here calling the radically strong version of punctuated equilibrium.

As rightly pointed out by Olivier Rieppel (1987), however, this is a worldview, a preconceived metaphysical option, rather than a scientific theory. It is not, and it does not need to be, anchored in empirical data. As a variant of the strong version of punctuated equilibrium, it in fact flies in the face of whatever is known about the evolution of species. But the main problem with this worldview is much deeper. The principle of continuity from one historical state of an object to another is what ensures persistence of the object's identity through time and what thus allows for a causal explanation of the historical change. Were there no continuity, no gradual transition through intermediates, but only instantaneous flips between such different historical states of an object, they could as well be different objects causally independent from each other.

Consider, for example, the life of a certain Joseph Doe, whom we could meet only a few times—as a newborn baby, a toddler, a teenager, an adult, and a dying old man. His body would be completely different from one encounter to another, as would be his mentality. How could we ever imagine he is in fact the same person all the time, if not for the assumption of his continuous existence and transformation? We might replace this assumption with empirical knowledge if we were able to keep an eye on this Joe Doe throughout his whole life, but otherwise we would have to rely entirely upon our belief on continuity of the process of aging. Yet no process can be observed incessantly, least of all processes that extend beyond the time scale of everyday observations. This analogy may shed some light on the problems of punctuational change as envisaged by the radically strong version of punctuated equilibrium. It is no accident that Georg Hegel, the great nineteenth-century German philosopher upon whose ideas Marx and Engels constructed their punctuational worldview invoked by Gould, had begun with consideration of the continuity and identity of human person through time.

Given our inability to make more than a few observations separated from one another in time, we always have to choose whether we prefer to interpret them as representing a single continuous phenomenon or a number of discrete ones. We are entirely free to opt for one or the other interpretation, to accept either punctuations or continuity. But there is an intellectual cost to each solution. By opting for continuity, we arbitrarily restrict the potential field of explanations for the observed phenomena. By opting for discontinuity, in contrast, we rule out the possibility of causal explanations for historical phenomena. For, if there is no continuity between Joe Doe the teenager and Mr. Joseph Doe the businessman, developmental psychology becomes impossible because we have no reason to believe a teenager can ever grow into a businessman. Exactly the same is the case with evolutionary biology. If there is no continuity between the ancestral and the descendant species, there can be no science of evolution as the mechanism of species origination from another species because we cannot possibly think of these species as ancestors and descendants but only as unrelated entities.

It is very important to emphasize, however, that one may adhere to the principle of continuity while dealing with one realm of reality, but to simultaneously allow for discontinuity in another realm. In my view, this is the case with the neodarwinian paradigm, which proclaims continuity of ontogenetic development and evolution but which nonetheless allows for instantaneous death and extinction by extrinsic causes. It would be absurd for neodarwinians to deny that a stone may fall from a rooftop and kill a pedestrian on the sidewalk; and it would be equally absurd to deny that volcanic eruption on a small island in the middle of the ocean may ultimately exterminate some endemic insect species. Ontogenetic development of individual organisms and evolution of species, however, are always viewed by neodarwinians as being driven by the continuous interaction between intrinsic biological forces and extrinsic environment.

By contrast, the radically strong version of punctuated equilibrium extends the realm of discrete events of encompass the evolution of species and higher taxa. It has then to bear the cost of this metaphysical presumption. It may be capable of describing the pattern of of phenomena, but not of explaining it causally by identifying the process (Rieppel 1986).

7.5 The Moderate Version

Several writings by Gould and Eldredge (1977, Eldredge 1982, 1985, Gould 1982a, 1982b, 1983, 1985) suggest still another interpretation of punctuated equilibrium. It is the assertion that the evolutionary history of the vast majority of species consists mainly of a very long period of evolutionary stasis, when the species remains in homeostasis with its environment and essentially does not evolve, thus maintaining its identity and even individuality. This *moderate* version of punctuated equilibrium constitutes the starting point for some of the modern attempts to reach beyond the neodarwinian paradigm of evolution by expanding it into a hierarchical theory of selection, the cornerstone of the proposed hierarchical paradigm of evolution.

Long-term evolutionary stasis of species, however, simply cannot be tested in the fossil record. Paleontological data consist solely of a small sample of phenotypic traits—little more than morphology of the skeletal parts—which does not allow us to make any inference about changes in a species's genetic pool or even about changes of the frequency distribution of phenotypes in a phyletic lineage. The nonpreserved portion of the phenotype of each fossil species is so extensive that it may always undergo considerable evolutionary changes that remain undetectable by the paleontologist. What appears then to the paleontologist as a species in a complete evolutionary stasis may in fact represent a succession of fossil species or perhaps a whole cluster of species, a phylogenetic tree with a sizable number of branching points, or speciation events. Even disregarding the hopeless instances of sibling species, which can only be distinguished by their molecular biology or by behavioral observations but not by their morphology, just consider the vast number of animal species that differ among themselves by nonpreservable morphological features;

for example, snails, which differ in anatomy of their foot and digestive tract; or spiders, which differ in structure of their genitals. Just consider what all could evolve in a shallow-water marine filter-feeding clam whose shell morphology and muscle scar pattern—the only features preservable in the fossil record—remain the same over millions and millions of years: foot structure as adaptation to burrowing in various soft substrates, gill structure as adaptation to various kinds of food, physiology of locomotion, feeding, and breathing, etc. The claim that long-term evolutionary stasis is the norm in the evolution of species cannot be either confirmed or refuted. It is untestable.

The moderate version of punctuated equilibrium, however, can also be interpreted even more modestly, as proposing that many, perhaps even the majority of, species exhibit long-term phenotypic, morphologic stasis in many, perhaps even a great many, respects. Even the very limited goal of demonstrating stasis in single phenotypic features is fraught with statistical and stratigraphic pitfalls. First, stasis must of course be defined quantitatively, so that its occurrence could be corroborated or refuted in any particular case. In the absence of quantitative definition, discussion may go on forever. While discussing the pattern of change in single morphologic characters in the Jurassic ammonite *Kosmoceras,* David Raup and Rex Crick (1981) had to withhold their opinion on whether it is stasis or gradual evolution just because they could not compare the empirical pattern to any quantitative standard of stasis. A definition of stasis must obviously avoid equating stasis with persistence of a species's phenotype within the boundaries set by the taxonomic standards employed for the species recognition; such equation would automatically ensure that every species remains in stasis over its entire duration.

Russell Lande (1986) proposed that stasis be defined as oscillations of the mean value of a phenotypic character within two standard deviations of the original population's variation. This definition roughly coincides with the limits set by systematic zoologists to subspecies. It is nevertheless entirely arbitrary and highly dependent on the variation of the original population, which may be related to the population's local ecological context and evolutionary history. On Lande's criterion, a widely variable original population is more likely to be regarded as remaining in stasis than a population with relatively little variability, even though the actual extent of change may be exactly the same. Given such definition, furthermore, sample size becomes critical; for if samples are too small, the statistical null hypothesis that their means are indiscernible, or at least differ by less than two standard deviations, may be "too null" to be rejected (Levinton 1982). The very nature of statistical testing puts stasis, as a null hypothesis, in somehow privileged position, and if the data are poor, stasis may win, so to speak, by default. The impression of stasis in various fossil species can moreover be reinforced by the time-averaged nature of fossil assemblages, which lump together individuals that actually lived centuries or millennia apart. Consequently, true phenotypic differences between populations in a temporal succession are artificially diminished by the composite nature of samples.

Very few cases of morphological stasis have actually been empirically demonstrated according to quantitative criteria, but this is not to deny the existence and even commonness, of such patterns in the fossil record. A Paleocene fossil (almost 60 million years old) of the clawed frog *Xenopus* is essentially indistinguishable by its bone morphology from living species (Estes 1975). A Late Cretaceous, 70-million-year-old, bryozoan fossil is indiscernible from the living species *Nellia tenella* (Winston and Cheetham 1984). Dozens of living bivalve species have their shell morphologies identical to those encountered among Early to Middle Miocene, 20-million-year-old, fossils (e.g., Hoffman and Szubzda-Studencka 1982). The evolutionary rates calculated for shell shape features in several bivalve species by Stanley and Yang (1987) who compared Pliocene forms to their presumed living descendants are so low they are hardly discernible from zero. Such morphological stasis is particularly striking in the light of the high tempo of molecular genetic changes. As pointed out by Thomas Schopf (1981), the genome is virtually constantly in flux. Morphological stasis well may cooccur with profound evolutionary change. The protozoan genus *Tetrahymena*, for example, includes a number of species that are morphologically so close to each other that they were traditionally regarded as having diverged only very recently; yet they are separated by vast molecular differences (Williams 1984). Similarly, the plethodontid salamanders studied by David Wake and his coworkers (1983) underwent much molecular divergence in spite of very little morphological change.

There is no way to estimate quantitatively the relative frequency of such phenotypic stasis over the geological time scale, simply because it is impossible to draw a random sample of phyletic lineages in which the pattern of morphological change through time could be empirically determined. Nevertheless, the widespread incidence of such long-term morphological stasis is widely acclaimed not only by the advocates of punctuated equilibrium, but also by its most ardent opponents. What is at issue, however, in the current arguments on morphological stasis, is its mechanism and its significance for evolutionary theory.

The *enthusiastic* variant of this moderate version of punctuated equilibrium—as represented, for instance, by Pere Alberch (1980) and Stephen Jay Gould (1980, 1982b)—holds that the morphological stasis observed in various species is primarily caused by developmental constraints imposed upon the potential of natural selection and other evolutionary forces by the intrinsic properties of the developmental system. A developmental constraint is here understood, following the consensus reached by the participants of a recent conference on the relationship between development and evolution (Maynard Smith et al. 1985), as a bias in the appearance of phenotypic variants in a species, where the bias has no relation whatsoever to the species' adaptive requirements. The enthusiastic variant portrays stasis as a result of the developmental system's resistance to change, as due to the development's homeostatic properties which allow solely for minor adjustments but not for any major shifts. This concept, then, gives great weight to the role developmental constraints, as compared to evolutionary forces, play in evolution. It calls therefore for a study of the ways developmental systems can be transformed from one state into another, for seeking the laws determining the patterns of origination and transformation of organic forms. It suggests a fundamental reorientation of the research programs intended to explain the patterns of morphological change in the history of life on the Earth. In this sense, the enthusiastic variant goes beyond the traditional understanding of the neodarwinian paradigm.

There can be little doubt that developmental constraints do play a role in shaping the pattern of phenotypic variation. The problem is only how to distinguish between developmental and selective constraints while seeking an explanation for the existence of certain organic forms or nonexistence of some other conceivable forms. This is a very difficult task, which can be successfully accomplished only under rather favorable circumstances. John Maynard Smith and his codiscussants (1985) give a good example concerning the wood- and rock-boring bivalves of the family Pholadidae. The shells of all bivalves grow as two logarithmic spirals in mirror image of each other. Pholads, however, bore by a rotary motion producing a cylindrical hole, and they therefore need to have a circular shell outline in cross section presented to the substrate. They achieve this end by taking a rather special position of the body. This is a solution to their problem, an adaptation, but it is optimal only given the fundamental constraint of the shell growing as logarithmic spirals. It would certainly be more advantageous to these animals to have simply a cylindrical shell. Yet this possibility is ruled out by the pattern of ontogenetic development.

There are many more examples of this kind, indicating the role of developmental constraints. But what all these examples demonstrate is merely that some limitations exist that cannot be, or at least cannot be easily, broken down by evolutionary forces. Whether such limitations do indeed induce morphological stasis, as asserted by the enthusiastic variant, is a completely different question. It is a question that cannot as yet be ultimately answered, simply because the evidence is unavailable. There is at present no data to suggest that the morphological stasis observed in any particular species—like the frog *Xenopus*, the bryozoan *Nellia tenella*, or the bivalves referred to above—is caused by developmental constraints that counteract selection and other evolutionary forces. There is no counterevidence either. It appears to me, however, that the very occurrence of extensive intraspecific variation in morphological characteristics indicates that such developmental constraints cannot provide an adequate explanation for morphological stasis at the species level.

The enthusiastic variant, however, does not necessarily follow from acceptance of the commonness of morphological stasis. The *skeptical* variant of the moderate version of punctuated equilibrium accepts the wide occurrence of phenotypic stasis but attributes it to a variety of biological mechanisms that are all fully compatible with the neodarwinian paradigm, though not easily discernible from each other in any particular problem situation.

Certainly, the simplest way to maintain stasis is by migration of populations and species that can track the habitats to which they have become adapted. During glaciation, for example, the seawater along the coasts of North America become cooler; it was as if marine habitats with particular water temperatures had "migrated" southward, except for the warmest-water, southernmost habitats which became narrower or even completely squeezed out. Hence, the benthic invertebrate faunas inhabiting the shelf areas also migrated southward, thus avoiding the new selective pressures rather than acquiring new adaptations. For as long as the habitat of a species does not undergo destruction but only a shift in space, the environmental change may not pose any challenge to be met by evolution. To be more precise, populations and faunas of course do not migrate as entities; individual organisms do. Under such circumstances, however, the fittest individuals are those that migrate in the appropriate direction, whereas the individuals that remain in place are most likely to survive in decreasing numbers because of competition with increasing numbers of better adapted immigrants from other species. The survival of migrants is what ensures stasis of the species.

According to Brian Charlesworth and coauthors (1982), the most powerful, and also the most standard, neodarwinian mechanism to account for morphological stasis is stabilizing selection, which is the kind of natural selection capable of weeding out extreme, and hence generally less fit, phenotypes. Given that at least certain aspects of the environment remain approximately constant over very long time intervals, it is entirely understandable that certain phenotypic traits will be consistently selected for at the early, relatively short-term stage of species evolution toward an adaptive peak in a new environmental situation the species encountered. As the species acquires new adaptations, however, and thus approaches an adaptive peak under given ecological conditions, the individuals that have phenotypes closest to the population mean become the fittest; hence, new mutant variants become unlikely to establish themselves. The early directional selection, pushing the species in a particular adaptive direction, is thus replaced with stabilizing selection, which then maintains the species in stasis until a change takes place in some relevant ecological parameters. The phrase "relevant ecological parameters" is crucial here; for, although a constancy in the totality of ecological conditions is virtually impossible in any given environmental context, the particular features of the environment to which organisms have morphological adaptations preservable in the fossil record may well persist for very long periods of time.

As pointed out by C. J. Barnard (1984), however, stabilizing selection as the mechanism of stasis implies that environment preexists the organisms that adapt to fit it. In reality, though, organisms also alter their environment by acting upon both its physical characteristics and its other inhabitants. They construct shelters, release metabolic waste, feed upon, and fall prey to, other organisms that did not previously feel their existence. These alterations, in turn, are likely to affect the organisms by resulting new selective pressures. Interdependence between organisms and their effective environment thus arises. Feedback loops in such a coevolutionary relationship may eventually lead to a stable equilibrium, and hence stasis, because each change in a species or population would most likely be counteracted by a variety of selective pressures coming from the environment.

Still another potential mechanism of morphological stasis refers to some intrinsic properties of the ontogenetic development of organisms. Ivan Schmalhausen (1949) and C. H. Waddington (1957) emphasized that organisms often have the ability to achieve the same adult phenotype in spite of considerable genetic changes and under a wide variety of environmental conditions. This ability of the developmental system to follow the normal developmental pathway, but also its inability to enter it if the genetic or environmental conditions are too far off the norm, has been termed "developmental canalization." As pointed out both by Schmalhausen and Waddington, such canalization is often of considerable selective advantage to the organism because it acts to minimize the adverse effects of disturbance on the course of developmental processes. This fundamental property of individual development permits genetic variation to accumulate in a population over time, without being expressed in the phenotype and hence without being perceived by natural selection. Under environmental stress, however, the canalization may break down, the accumulated genetic variation may lead to phenotypic variation, and natural selection thus has more factors to act upon. As a result, the period of morphological stasis may be over. The opposite to developmental canalization is the phenomenon of developmental plasticity, with a single genotype producing a wide variety of phenotypes, depending on environmental conditions. But the point of the matter is that species vary in the strength of their canalization and plasticity, and that these properties clearly can be adaptations, subject to natural selection because they contribute to fitness. Stephen Stearns (1982), myself (Hoffman 1982), and David Wake and coworkers (1983) independently proposed that such developmental canalization can promote morphological stasis, at least in some characteristics of the organism.

The skeptical variant of the moderate version of punctuated equilibrium thus differs from the enthusiastic variant in that it perceives morphological stasis as a byproduct of the action of natural selection and other evolutionary forces, as an accident of life. It is very difficult to corroborate one or another of the several mechanisms it envisages as the causal explanation for morphological stasis in any particular case from the fossil record, but the mechanisms are certainly plausible. In fact, at least some circumstantial paleontological evidence can be found in the common occurrence of phenotypic variation increasing during periods of accelerated evolution and decreasing during periods of stasis (Hoffman 1982, Williamson 1987).

In its skeptical variant, the moderate version of punctuated equilibrium thus falls well within the neodarwinian paradigm of evolution. And although it may have drawn more attention to morphological stasis, a phenomenon little studied by evolutionary biologists, it does not lead to any new insights. It actually borders on triviality in that it very closely approaches what I described above as the weak version of punctuated equilibrium. The only difference lies in the assertion that morphological stasis rather than change prevails in the history of life on the Earth and must therefore be explained in the first place. This assertion, however, can hardly be regarded as anything more than a minor shift in emphasis. It certainly does not lead to formulation of any new laws that have to be taken into consideration while explaining historical biological phenomena.

7.6   Final Assessment

After fifteen years, the debate on punctuated equilibrium may well be closed. Contrary to the triumphant proclamations of Gould and Eldredge (1986), however, punctuated equilibrium has not been ultimately established as a truth that now everyone claims to have always known and accepted. It is not part and parcel of the evolutionary theory, indispensable for causal explanation of the course of the evolution of species. Punctuated equilibrium has a number of versions, and depending on the version accepted its assessment must also be different.

The weak version of punctuated equilibrium is entirely trivial, that is, known ever since Darwin and accepted by the evolutionary theory at least since the advent of the Modern Synthesis in the 1930s and 1940s. The moderate version is untestable in its original formulation, pertaining to the true evolutionary stasis of species. When understood in purely morphological terms, as concerning only certain morphological features of the organism, those preservable in the fossil record, this version is either trivial (the skeptical variant) or seemingly wrong or at least unsupported by any evidence whatsoever (the enthusiastic variant). The mildly strong version is plainly false, and the radically strong version is a metaphysical rather than scientific statement and hence untestable.

Punctuated equilibrium cannot therefore force the evolutionary biologist to rethink, and go beyond, the neodarwinian paradigm. It does not justify a causal distinction between micro- and macroevolution; it does not indicate the operation of a separate class of macroevolutionary processes. It cannot serve as an argument in the current discussion on macroevolution—in spite of the fact that it largely triggered this debate.

Punctuated equilibrium is even less a milestone in the process of a fundamental change in the modern perspective on the world. The radically strong version of punctuated equilibrium portrays itself as a rebellion against the Western paradigm of continuity. It explicitly places itself within the catastrophist tradition of thought, with its emphasis on abrupt, qualitative changes. Gould (1984) took this stance with a particular pointedness. A prominent literary critic took this interpretation even further and depicted the relationship of the prose of Jorge Luis Borges and his fellow postmodernists to the classic novel in a close analogy to punctuated equilibrium as a revolution against the nineteenth-century paradigm of science (Levine 1986). Both the weak and the moderate versions of punctuated equilibrium, however, are not at all revolutionary and in fact do not speak for abruptness of change. The mildly strong version, on the other hand, clearly is an aborted revolution, dismissed by the evidence and, consequently, abandoned by its authors themselves. The radically strong version, in its turn, does not stem from any analysis of the empirical reality but, to the contrary, is a metaphysical *a priori* that may or may not be accepted regardless of what this reality is like. At this point, however, nothing suggests it to be particularly productive as a starting point for explanation of historical biological phenomena. This is not to say that the paradigm of continuity must win overall—either in evolutionary biology, or in science as a whole, or on a broader cultural scale—but only that the scientific meaning and validity of punctuated equilibrium is irrelevant to this issue.

As a problem in evolutionary biology and paleontology, punctuated equilibrium by itself has by now reached the stage where there is nothing to debate any more. The discussion that has brought us to this stage, however, has two undeniably valuable merits. On the one hand, it stimulated a lot of exciting empirical research that has resulted in some fascinating data; and it also enforced much higher standards of paleontological work on the tempo and mode of evolution. On the other hand, it helped to bring the concept of species as logical individuals and the theory of species selection to the focus of evolutionary biology. Each of these two ideas, in turn, has triggered as much of a vigorous debate as did punctuated equilibrium itself. The latter has thus significantly contributed to maintain evolutionary biology in the healthy state of continual flux that makes it so fascinating a field for empirical and theoretical research as well as for historical and philosophical analysis.

8 Species as Individuals: A Philosophical Problem

The concept that biological species are individuals rather than classes or anything else has been repeatedly portrayed as a revolution in the understanding of evolutionary theory, one that enforces a rethinking of the Darwinian perspective on evolution (Eldredge 1985) and a reconsideration of the methodological structure of biological science (Rosenberg 1985). It represents, however, a stance in the old philosophical debate on the logical status of species.

Plato and Aristotle began this debate more than two millennia ago, in ancient Greece, when they discussed the nature and existence of such abstract ideas as stone, or eagle, or knife. For Plato, these general ideas had a real existence somewhere in the universe, and all particular stones, eagles, and knives were their exemplifications. Through Aristotle's enormous impact on all natural sciences, the concept of particular organisms being such a member of a general category of beings, a given organic species, and hence of each organism having the diagnostic features of its species and thus resembling its abstract type, became established in preevolutionary biology. This concept is the foundation of Karl Linné's *Systema Naturae* (1735). Yet general and abstract categories of beings, or classes, exist forever in the same form, no matter if there are in the universe any of their particular, real elements. Logical classes are immutable and eternal.

William Occam, however, a brilliant English philosopher of the Middle Ages, argued that there was no evidence of and no need whatsoever for the existence of such general ideas. Why not accept that all particular beings are just what they are—individual stones, eagles, or knives—classified by human mind into more or less arbitrary groups for the sake of convenience and interpersonal communication? There is no abstract stone, abstract eagle, or abstract knife but only this or that particular one, which you and I can look upon, touch, and grasp. According to this view, classes of elements have no real existence; only individuals exist, although these individuals may be composite in nature. And individuals, of course, can come into being, change, and be destroyed. When I observe them over years and decades, this piece of sandstone in my staircase wears down, the eagle in Warsaw Zoo gets older, and my old Boy Scout knife becomes rusted. If I crushed this piece of sandstone, killed this eagle, or melted this knife, they would all cease to exist and never reappear again.

This brief account of the philosophical debate is extremely simplified, to say the least, but it may help to set the stage for considering the problem of species as individuals. For what are, then, biological species? We are here on the horns of the Darwinian dilemma. The Aristotelian and Linnean species are real but abstract and eternal categories; the Occamian species are unreal, arbitrary collections of individuals. The former cannot evolve; the latter dissolve each species into a set of individual beings, each organism with its own properties and life history, and cannot be a meaningful object of study. Neither concept is satisfactory to an evolutionary biologist.

Darwin did not tackle this issue explicitly. Neither did his successors, evolutionary biologists, and the problem lingered until very recently at the interface between philosophy and biology. Currently, Michael Ghiselin (1974, 1987) and David Hull (1976) and a host of other philosophers of biology argue that given that all species show intraspecific variability, both genetic and phenotypic, and given also that all species are capable of evolving, even though the change accomplished in this way may not be particularly striking, species cannot be regarded as classes of individual organisms. There exist in the organic world no essential features that an individual organism would have to possess in order to belong to any particular species. Species are historical entities, localized both in space and in time. Each has its beginning (the amount of speciation), its evolutionary history (whether it is stasis or change), and its ending (the moment of extinction); no species reappears once it has become extinct. Moreover, all biological criteria employed to define and recognize species refer to such properties of organisms and populations as offspring fertility, reproductive isolation from other populations, and sharing in a common mate recognition system. They thus involve characteristics that reflect biological interactions among organisms and the resulting cohesion of species as a whole.

Briefly, species are neither immutable, nor eternal. Yet that is exactly what logical classes are. Organisms belong to a species because of their relationship to some other organisms, and not because of their having the essence of this species. Yet the inverse is true of logical classes. Thus, Ghiselin and Hull argue, species are not classes. The class of atoms of hydrogen is a typical logical class. Whichever atom in the universe has one single proton in its nucleus belongs to this class. Having one proton in the atomic nucleus is the essence of being an atom of hydrogen. Whether an atom of hydrogen is bound to another one and thus forms a molecule, and whether it makes part of a chemical compound, it is a member of the class of atoms of hydrogen. The essence of being an atom of hydrogen does not depend on its interaction with other atoms and does not change through time. In the sense of its abstract essence, the class of atoms of hydrogen persists eternally unchanged, whether any atoms of hydrogen actually exist in the universe. By contrast, individual members of the class, that is, individual atoms of hydrogen, do change a variety of their attributes through time and in interaction with other objects. They move in space, acquire or lose energy, get involved in chemical reactions, etc. They may even be formed by radioactive decay of heavier atoms or destroyed forever by thermonuclear reactions. Each atom of hydrogen has its own spatiotemporal location and its own history—its own identity. Ghiselin and Hull argue that if species are not logical classes, they must be logical individuals. And indeed species seem to have many attributes prerequisite for such interpretation of their logical status.

This whole argument is entirely independent, both logically and historically, of punctuated equilibrium and the claims for a causal distinctness of macroevolution from microevolution. An advocate of punctuated equilibrium might in fact argue both ways. On the one hand, he or she might point to evolutionary stasis of species as the norm in the history of life on the Earth. But if species do not evolve but only persist unchanged throughout their duration, they each have a set of invariable features—the essence of what it is that determines the membership in any given species. Therefore, species are to be regarded as classes. While advocating what I described above as the enthusiastic variant of the moderate version of punctuated equilibrium, Stephen Jay Gould (1983) in fact explicitly espouses the view that species' essence is more fundamental than their variation; and he describes the structure of developmental programs and genetic systems as the essential properties of individual organisms that determine their membership in a species. Given the radically strong version of punctuated equilibrium, this view appears to be even more logical because the inevitable absence of continuity, or intermediates, in transition between species suggests that each species has an essence that cannot be changed but only undergo a saltational switch, in a single step, from existence to nonexistence in the real world.

On the other hand, however, both the moderate and the strong versions of punctuated equilibrium emphasize the sharpness of the beginning as well as the ending of the evolutionary history of each species. They thus emphasize the spatiotemporal restrictedness of species, their real and historical rather than abstract nature. For Niles Eldredge (1982, 1985) and Elisabeth Vrba (1985b; Vrba and Eldredge 1984)—but also for Gould (1982c; Vrba and Gould 1986)—this corollary of punctuated equilibrium reinforces the position taken by Ghiselin and Hull. Species are to be regarded as logical individuals, not classes.

This point of view appears particularly attractive, though not particularly illuminating, to the biologist, simply because the interpretation of species as having essences—the essentialist, or typological, concept of species—has become obviously unacceptable at least since everyone agreed that species show variability, evolve, and originate by evolution. Ernst Mayr (1987) and Ledyard Stebbins (1987), two of the founding fathers of the neodarwinian paradigm, noted that for 50 years or so no biologist has thought or argued in terms of species as logical classes, even though several philosophers apparently did. From the modern biological, post-Darwinian perspective, species are not groups of individual organisms put together by the taxonomist because they share the essential features of an ideal type; nor are they merely arbitrary collections of individuals grouped by the taxonomist on the basis of their similarity in whatever aspects he or she may consider relevant. Species are evolving populations of individual organisms that at least potentially interact with each other and share in the common genetic pool. For the post-Darwinian biologist, species certainly are neither the Linnean classes, nor the Occamian artefacts of human mind. Their logical status might therefore be that of individuals.

The concept of species as individuals appears also particularly attractive to the advocate of macroevolution as a separate category of evolutionary processes because logical individuals can be subject to selection whereas classes cannot. As argued by David Hull (1980), individuality is in fact a necessary condition for an entity to be a unit of selection. For selection is a process resulting from variation, multiplication, and interaction with environment; since these three concepts are nonsensical with respect to classes, classes cannot be subject to selection. Selection, furthermore, is a process leading to a change in frequency distribution of entities; since classes are eternal, their frequency distribution must remain constant. Individuals are made up of what can be selected, either by accidental vagaries of the environment or by selection for or against some of their specific traits. Therefore, given that species are individuals, one can speak of species selection, and not only of natural selection within populations and species. Species can then be regarded as independent actors in the ecological and evolutionary theater, their fate and its consequences for the biosphere being determined by historical contingencies as well as by some biological, macroevolutionary laws that are irreducible to those of the genetical theory of evolution. This corollary has been taken up by Eldredge, Gould, and Vrba who made it a part and parcel of the current drive among paleontologists and some other evolutionary biologists to establish macroevolution not only as a category of phenomena but also as a class of processes, different from the microevolutionary ones envisaged by the neodarwinian paradigm.

Individuality alone, however, does not guarantee a capability of being subject to selection. As David Hull pointed out in the article in which he argued that only individuals and not classes can be subject to selection (Hull 1980), individuality is a necessary but not sufficient condition for an entity to be a unit of selection. This is in fact quite obvious because some perfect logical individuals do not meet the other conditions for selection. They have no variation and do not multiply. The biosphere, for instance, is an excellent example of an individual—with its spatiotemporal location, its history, and its internal cohesion—but it is unique, at least for now, and to speak of its being a unit of selection would certainly mean that the word "selection" is being employed in a completely different sense than usual in evolutionary biology. Atoms of hydrogen also are individuals, but they can hardly be conceived of as being subject to selection. Thus, even if species are individuals rather than classes, this still does not imply that one could reasonably speak of species selection as a causally distinct, macroevolutionary process. The implications of the concept of species as individuals for macroevolution are not as dramatic and far-reaching as they are often presented to be. What is indeed crucial in this context is that species are not classes of individuals whose membership in a given class is determined by immutable and eternal essences. But this conclusion is, of course, nothing new.

To conceive of species as logical individuals, moreover, does not solve the problem of essentialism in evolutionary biology. For, if species are individuals, then the question immediately arises, how do species maintain their individuality—or better, their identity—through time? Olivier Rieppel (1986) observes in his review and critique of the argument for species being individuals that two contrasting answers can be given to this question. Either one attributes the individuality of a species to its material continuity over its entire geographic range and throughout its duration in time, from its very origination as a logical individual to its ultimate demise; or one postulates such identity through time as a result of the process of mental abstraction, an inference based on some empirical observations but without certainty as to its nonarbitrariness. How else could we determine if the humankind we are currently observing on the Earth and the people who lived 30,000 years ago at Cro-Magnon, not to speak of the almost 2 million years old *Homo habilis* from Olduvai Gorge, represent different historical stages of one single species-individual or rather several different species? Either we have something real in common, or we only think we belong to the same species as the Magdalenian inhabitants of Cro-Magnon.

The former answer to this question, however, is essentialist, because it portrays species as having some actual, material features common to all individual organisms, or at least all populations, of a given species. These common characteristics may refer to some part of the genome, or the genetic pool, or whatever else; but given the constant flux of individual organisms and populations through time, the material continuity of species must depend on something deeper or more esoteric than direct empirical observation. These material features that ensure species continuity through time constitute then the individual essence of species. Yet essences can only come into being and be destroyed but not change in a continuous fashion. This is why the concept of species as individuals is compatible with the strong version of punctuated equilibrium.

In turn, the other answer to the question about the sources of species individuality deprives species of their historical reality and represents them as mere constructs of human mind. This corollary is unacceptable as a basis for the concept of species being the stuff of macroevolution, and it is therefore no wonder that Gould (1983) so explicitly endorses the essentialist interpretation of species. In fact, the concept of species as purely mental abstractions is also unacceptable to the vast majority of neodarwinians—as personified, for instance, by Ernst Mayr (1987)—if only because even a cursory observation of the living biosphere clearly indicates that species very often, though perhaps not always, can be unequivocally delimited and recognized. For the biologist, species are real.

No wonder that philosophers vehemently argue about the logical status of organic species. For if their conclusions are to conform to the biological usage, if they are to help to remove confusion, misinterpretation, and equivocality rather than to confer a totally new meaning to the term "species" and thus to produce even further confusion—species must not be conceived of either having an essence or being purely mental abstractions, without a historical reality. Evidently, the conceptual opposition between classes and individuals is not exhaustive; it does not adequately describe the whole reality. The world is richer in kinds of entities than just these two logical categories. In particular, some other categories should be applied to biological species.

This assertion is not at all new either among biologists or among philosophers, though I may justify it here differently. The biologist Leigh Van Valen (1976) proposed to consider species as individualistic classes, where the term "individualistic class" designates an entity that functions as a class in one context but as an individual in another. A person is an individual, but a person is also a class of its cells, with the genome of each cell determining its membership in the class, hence being its essence. Ernst Mayr (1987) and Ledyard Stebbins (1987) regard species not as individuals but as populations and systems, respectively, in order to emphasize their composite and internally structured nature. The philosopher Philip Kitcher (1988), in turn, suggests that species be regarded as sets—perhaps fuzzy sets because it is quite often unclear whether an individual organism should be assigned to one species or another—defined by a variety of attributes, depending on the biological context. This is a pluralistic concept of species. A set may be defined arbitrarily, and this fact has prompted some severe criticisms of Kitcher's position. Yet not all sets are defined arbitrarily. The dilemma is how to define species in a nonarbitrary way and yet have them correspond to the biological reality. This problem, however, is no more acute for species as sets than it is for species as classes (what kind of logical classes without essence are species?) and species as individuals (what kind of logical individuals are species in contrast to, say, individual organisms?).

The question concerning the logical status of species demands a satisfactory answer. It is a problem to be solved. But it is a problem for philosophy not biology. For the whole issue is of little relevance to the biologist. The biologist does not draw any conclusions about the investigated species from a philosophical consideration of their logical status, but he or she focuses on the biology of these species. The philosopher may on this basis discuss the nature of species and attempt some generalizations about their logical status. But the biologist knows that species are neither eternal, nor immutable. They are historical entities, localized both in space and in time, equipped with some cohesion and capable of evolving; and no single criterion has ever been given that might permit one to define clearly and unequivocally species at all possible kinds. The biologist knows that no matter whether species are called classes, or sets, or individuals, or whatever other logical terms may still be invented, their biological status—both in systematic or ecological practice and in evolutionary theory—must reflect these biological properties.

9 Species Selection: An Explanation in Search of Phenomena

9.1 The Unit(s) of Selection

On a wide variety of occasions, Niles Eldredge, Stephen Jay Gould, and Steven Stanley identified the theory of species selection as the most, perhaps even the only, important corollary of punctuated equilibrium. John Turner (1986), one of their most virulent critics, agrees, although he still maintains that punctuated equilibrium is all wrong. Many other neodarwinian evolutionists, however, attribute at best a marginal significance to species selection (for example, Dawkins 1982, Maynard Smith 1983), whereas Max Hecht and I add further that its relationship to punctuated equilibrium is questionable to say the least (Hecht and Hoffman 1986). This situation is at least equally confusing as it is in the case of punctuated equilibrium itself.

The term "species selection" refers to a biological process of selection operating upon species as units. It thus implies a process at the level of species being superimposed on the process of natural selection within populations and species, as envisaged by the neodarwinian paradigm of evolution. The theory of species selection is therefore a step beyond the neodarwinian paradigm, toward a paradigm of evolution that would envisage at least a twofold hierarchy of processes of selection and a certain interaction between its focal levels. Hence, it is widely regarded as the cornerstone of the new hierarchical paradigm of evolutionary biology, which is allegedly about to replace the traditional neodarwinian one by subsuming it within a more expanded conceptual framework. The neodarwinian paradigm is then viewed as focusing on only a subset of evolutionary processes and thus oversimplifying the actual reality of life on the Earth (Eldredge 1985, Gould 1985, Salthe 1985).

The kinds of biological entities that can be, and actually are, units of selection—and hence, the levels of the biological hierarchy of organization that can be, and actually are, the focal levels of selection— are subjects of a longstanding and very intense discussion among evolutionary biologists. Natural selection is generally understood in biology as selection *of* some objects due to selection *for* some of their characteristic features. The objects must multiply and show a variation in inheritable features that interact with environment, causing differences in the individual rates of multiplication. Given these conditions, the frequency distribution of the considered objects will change through time, and if the objects are biological entities, the causal process responsible for such change will be organic evolution driven by the evolutionary force of natural selection.

Evolution through natural selection thus means that something increases in frequency among a group of comparable biological entities, at the same level of the biological hierarchy, because that something has some inheritable features good for it, features that enhance its ability to compete for scarce resources and improve its chances in the incessant struggle for life and maximally successful reproduction. But what is that "something" that can increase in frequency by this way? There is obviously a whole hierarchy of biological entities (from individual genes to organisms, populations, species, monophyletic groups of taxa, and even the entire biosphere), and a pattern of temporal change in frequency distribution of such entities can be detected and described at each of these levels. That is, except for the grandest one, the biosphere, since there exists only one entity of this sort and it can change its structure but not frequency distribution. For example, the sickle-cell allele in a hemoglobin locus in humans may decrease in frequency in West African human populations due to the falling importance of malaria as the cause of selective pressure. Small elephants may increase in frequency on an island relative to the larger-sized members of the same population. Baboon bands with well-expressed and rigid dominance hierarchy among males may become with time more common in a certain species, at the expense of more loosely organized gangs that are more vulnerable to outbreaks of intrapopulation violence. Marine gastropod species with long-living planktic larvae, capable of floating very large distances, may decrease in frequency in a certain genus or family of gastropods, whereas species with larvae settling on the substrate immediately after hatching from the egg may become more and more numerous. Analogous patterns might also occur among marine gastropod genera or families within an order. But does it mean that all these patterns of evolution are caused by selection, which indeed operates at each of these levels? Why would then Darwin and the neodarwinians, who have obviously been fully aware of the elaborate hierarchy of nature (just recall what Bernhard Rensch and George Gaylord Simpson wrote on macroevolution) choose a much narrower view of evolution through selection?

Patterns of change through time can certainly be described at all these levels and each explained by a process of evolution. But it does not necessarily follow from this observation that evolution at each of these phenomenological levels of the biological hierarchy, as Niles Eldredge (1982) aptly put it, must be viewed as driven by the forces of selection specific to these particular levels. Whether a certain kind of biological entity can be subject to selection is a legitimate and important question, though relatively uncontroversial one since the vast majority of evolutionary biologists agree that a wide variety of categories of biological entities can indeed be validly regarded in this way. But whether a certain kind of biological entity can be subject to selection due to selection for its traits is an altogether different question. And it is this latter question that is at issue in the arguments on species selection.

What can be a unit of selection, or a target of selection as Ernst Mayr (1986) would prefer to call it in order to emphasize that such an entity should not only be subject to passive selection of but also be actively selected for its features? Back in the 1930s and 1940s, when the neodarwinian paradigm was formed, the traditional Darwinian focus on the individual organism as the unit of selection was taken for granted. According to this view, selection always operates on individual organisms and promotes all their features that contribute to increase the fraction of the population that their progeny will constitute in future generations, that is, given the persistence of the current environmental framework.

This classic concept of natural selection at the individual level is entirely compatible with the commonsense perspective on the ineradicable drive of organisms toward procreation, which is in common to all living beings, but it also allows for a variety of survival and reproductive strategies depending on both the environment and the biological constitution of organisms. It predicts, for instance, that in rapidly and unpredictably changing environments, under conditions of commonly but irregularly occurring catastrophic mortality (for example, in a coastal lagoon subject to storms and heavy rainfall), organisms will tend to grow rapidly, achieve early maturity, and produce at once vast amounts of eggs, generally poorly equipped in yolk and hence energetically cheap. Under stable and predictable, resource-poor environmental conditions, where food is scarce, competition intense, and populations are close to the maximum numbers sustainable by the environment (for example, on the bottom of deep oceans or in deep caves), organisms will tend to grow more slowly, delay maturation, and reproduce several times, giving birth to only a few relatively large and strong offspring which can do better as competitors.

This view of natural selection, however, encounters a serious problem while dealing with apparently altruistic behaviors of many organisms. Why should a worker bee, wasp, or ant take care of her sisters and brothers produced in large numbers by the queen, rather than go about having and rearing children of her own? Why should a bird give an alarm call, and thus alert the whole flock to the predator by taking the risk of attracting attention to itself, rather than hide itself quietly and wait until the predator goes away with one or more competitors of the bird in the claws? Provided that one's own maximally successful reproduction is the main goal inevitably pursued by all organisms evolving through natural selection—simply because organisms that do not pursue it at all or pursue it with relatively little success sooner or later disappear from the Earth without leaving any progeny—such behaviors are hardly understandable.

Many biologists resorted therefore to the reasoning that selection may also operate for the good of the population or species. In their perspective, a worker ant or bee refrains from having children because a division of labor between the queen and the workers is more advantageous to the population; it offers more chances for its survival in competition with other populations and species. A bird voices alarm because such behavior helps the population survive the risks and dangers in the world full of predators. According to this view, natural selection brings about such behaviors because their benefits to the population outweigh the costs to the individual. The evolutionary force invoked to explain the occurrence of such apparently altruistic behaviors by a reference to their benefit for the population or species has come to be known as group selection.

The group-selectionist position was particularly strongly advocated by the great Scottish zoologist V. C. Wynne-Edwards (1962), who argued, for example, that animals might have a smaller number of offspring than physiologically possible, in order to ensure survival of the population which could otherwise go extinct due to the self-induced shortage of resources. Curiously, almost the same argument was also made by Trofim Lysenko, the infamous dictator of the Soviet biology under Stalin. He ordered that oak seedlings in Central Asian arid lands and semi-deserts be planted in very dense clusters, since he hoped they would cooperate with each other for the sake of the population, or each particular cluster, and thus help to grow patches of woods.

It should come as no surprise, however, that Lysenko's little oaks did not cooperate but rather intensely competed with each other, and the project was a failure. Group selection in this extreme form is very unlikely to operate in nature. John Maynard Smith (1976) made a very simple but powerful argument to this effect. Consider a population the members of which have been conditioned by group selection to forgo some potential advantages to their individual offspring and to cooperate with each other for the sake of their common welfare. If such a population is invaded by an immigrant that still acts exclusively for the sake of its own offspring—to increase as much as possible the fraction of the population its progeny will constitute in future generations—it shall certainly do better in reproducing within this population than its more altruistic members. Provided therefore that these behaviors have a genetic background and are inheritable, an assumption shared by the theory of group selection, the selfish individuals will soon outcompete the altruists in the population. In the nonhuman world, altruistic behaviors should therefore occur at best temporarily, until selfishness takes over again.

Thus, group selection cannot be the solution to the dilemma that the phenomena of apparent altruism pose to the genetic theory of evolutionary forces. Another solution is needed. Such a solution has been postulated by William Hamilton (1964) who suggested that natural selection at the individual level acts not so much to increase the frequency with which progeny of any given individual organism will be represented in future generations, as to increase the likelihood that copies of its genes will be perpetuated in maximum possible numbers. Yet full brothers and sisters share a half of their genes on the average; half brothers and half sisters have a fourth of their genes in common on the average; and first cousins are most likely to share one-eighth of their genes. Hence, an act of apparent altruism may in fact be a purely selfish deed, because even a sacrifice for more than two sibs, more than four half sibs, more than eight first cousins, and so on, may contribute more to improve the chances for propagation of the copies of one's own genes than selfishness could possibly bring about. Genetically speaking, it may pay off to help one's kin. This is the basic tenet of the theory of kin selection. This theory, in its turn, is the foundation of sociobiology.

Kin selection may explain much of the apparently altruistic behaviors in nature and thus meet the challenge posed to the traditional Darwinian concept of natural selection operating solely at the individual level. But within the kin-selectionist conceptual framework, the individual organism is no longer the unit of selection. The evolutionary force of selection acts to increase the chances that the largest possible fraction of the individual genome will be multiplied and increase in frequency. It is the genetic material of each individual—or better, the sum total of its copies—which is subject to selection of. In this view of natural selection, it does not matter at all whether my own genes or the identical genes encountered in my twin sister or half brother will be copied and propagated; and it also does not matter at all which particular genes it will be. The individual organism thus dissolves as an entity to be selected, although selection of the genetic material is caused by selection for its phenotypic effects on the individual organism. Those genomes and their parts that lead to the most successful phenotypes will be promoted by natural selection. The reproductive success of individual organisms gives the criterion of which traits are to be selected for or against, but the genomes and their components are what increases or decreases in frequency in the population.

Technically, Hamilton's kin selection is very clearly defined. What is maximized by the process of evolution driven by this evolutionary force is the inclusive fitness of genomes, that is, the probability that the greatest proportion of their components—individual genes understood as those parts of the genome responsible for any given phenotypic traits—will spread in the form of as many copies as possible in future generations. The genetic material, or rather the information it carries, persists in spite of death of individual organisms; it multiplies and can undergo selection.

The concept of the unit of selection, which is inherent in the theory of kin selection, may nonetheless appear to be rather vague and confusing. For, what is it exactly that is subject to selection of? As soon as the individual genome gets involved in sexual reproduction, it dissolves and undergoes mixing and reshuffling along with the components of another genome. It does not multiply. Individual genome per se cannot be the unit of selection, although the traits it brings about in interaction with environment are what is selected for or against.

It is parts of a genome that are indeed multiplied and can be regarded as subject to selection of; more exactly, the genetic material in common to the kin is the target of selection. But how can we determine the limits of this genetic material within the individual genome? The process of gene recombination during sexual reproduction ensures that there never be certainty whether a given component of the parental genome will be transmitted to a particular son or daughter. Moreover, the vagaries of meiosis make all measures of genetic relatedness between various kin probabilistic; my daughter may have more of my genes in common with my son than with my other daughter. The concept of inclusive fitness, which is at the core of the theory of kin selection, is therefore very deeply probabilistic in its nature. It not only refers, as do all other measures of fitness, to the probability of reproductive success in the long run and as compared to other members of the population. It also takes into account the probability that an individual's kin is equipped with the theoretically expected proportion of its genome, and it also considers the probability that this shared proportion of the genome includes the particular gene whose phenotypic effects are selected for or against. It is correct to say that the genetic material is subject to selection of due to selection for its phenotypic effects, but the exact meaning of this phrase is hardly extractable from all those probabilities.

It is at this point that Richard Dawkins and his famous *Selfish Gene* (1976) enter the scene. For Dawkins, the gene is unambiguously the unit of selection. Individual genes exist in many identical copies in every population, and they multiply at variable rates depending on the reproductive success or failure of their carriers—individual organisms. This reproductive success or failure, in turn, depends on the interaction between the phenotypic effects of genes and the environment. What really counts in this context is, of course, the entire genome and its phenotypic reflection—for it is the whole organism that survives, reproduces, or dies—but particular traits, or the phenotypic effects of each particular gene, contribute their share to the total phenotype and thus have bearing on the reproductive success of the individual organism. The gene is here obviously defined not in exact molecular terms but rather as the genetic background of a phenotypic trait, no matter how much genetic material it entails in terms of the nucleotide bases of DNA strand. Thus, individual genes increase or decrease in frequency through time in a population, this change being brought about by adaptive advantages or disadvantages their phenotypic effects confer upon their carriers. The individual gene is therefore subject to selection of due to selection for its effects. Natural selection at the individual level operates upon genes. Individual, selfish genes are the real players in the evolutionary theater.

As put explicitly by Dawkins himself in his second book (Dawkins 1982), this concept of selfish gene is no more valid or correct as interpretation of the nature of the evolutionary force of selection than is Hamilton's theory of kin selection. Dawkins portrays the relationship between these two interpretations of natural selection and the biological reality as analogous to two different but equally true perceptions of the Necker Cube. The Necker Cube is a line drawing that gives impression of a cube. When watching the drawing, however, the mind flips every once in a while from the picture of one cube to that of another one, for the drawing is compatible with both of them.

This parallel holds for selfish gene and kin selection. Either concept explains the workings of natural selection, though neither does so without encountering a conceptual problem of its own. For the kin-selectionist position, the unit of selection dissolves into vagueness; for the gene-selectionist view, in its turn, the existence of individual organisms becomes a phenomenon that calls for explanation. Why should individual genes be grouped in large bunches and thus enter jointly the interaction with environment, which determines their evolutionary fate, rather than operate individually? Why should individual organisms contain more than one gene each? The vagaries of reshuffling and mixing during sexual reproduction are bound to cause some copies of otherwise very successful genes to perish due to their accidental linkage in a genome to less advantageous genes. This effect would be ruled out if each gene were operated individually and hence subject to selection exclusively on the basis of its own phenotypic effect. Yet genes cannot be separated from the genome. In other words, given the gene-selectionist framework, the organization of the genetic material into large and complex genomes that bring about whole organisms must also have been selected for.

To explain this phenomenon, Dawkins considers a world of individual genes, not yet grouped into bunches that could be regarded as whole genomes responsible for organization of whole organisms. Such a world of course cannot exist because genes as we know them now can only operate in the context of other genes, but it is useful to consider this presently impossible situation in order to find out what the consequences would be if it were in fact possible long, long ago. Dawkins points out that if a gene arose whose phenotypic effects in such a world of individual genes are advantageous for selection of some other gene whose phenotypic effects, in turn, are advantageous for the spread of the first gene, there will be selection for both these phenotypes and also for their coupling together, so that they be more likely to meet together and to operate more efficiently. This will be selection for gene organization into the simplest genome. By extension, this argument leads to reappreciation of the genome and its carrier—the phenotype of the individual organism. Thus, we come back to the other perception of the Necker Cube. We return to the perspective on natural selection as acting for propagation of the greatest portion of the genome, even if it meant a drop in the rate of propagation of one or another particular gene.

9.2 Population and Species as Benchmarks of Evolution

Exactly the same argument, however, can also be applied to groups of organisms, populations, and species that may then also emerge as cohesive biological entities. Insofar as such entities are capable of multiplication and possess some hereditary properties, they can change their frequency distribution through time and thus be subject to selection of. One can, for example, imagine that a certain flock of birds acquires an inheritable behavioral feature absent from all other flocks of this bird species, and that after a certain time migrants from the flock invade other flocks and colonize other habitats, spreading this feature. Thus, the number of flocks having this new feature increases through time. This increase may or may not correspond to an increase in the number of individual birds, but insofar as we are concerned with the number of flocks—in other words, insofar as we consider the flock as a benchmark of selection—we clearly deal in such a case with selection of flocks. Similarly, a wingless species may appear in a certain genus of grasshoppers that has thus far included solely winged organisms. If this new species subsequently gives rise to other wingless species that speciate further into ever new wingless forms, the proportion of wingless species in the genus may increase in the course of evolution, even though the proportion of wingless individuals in the genus may remain roughly constant. Again, insofar as we are interested in the number of species, or consider the species a benchmark of evolution, we may here speak of selection of a particular category of species within the given group of grasshoppers. One might even go still further and speak of selection of, say, tetrapod clades among the vertebrates, since the proportion of amphibian, reptile, bird, and mammalian clades within the Vertebrata has certainly increased in geological time. But can such groups of organisms, populations, species, or clades be true units of selection? Can they be subject to selection *of* due to selection *for* their specific traits? Can we validly speak of natural selection operating at the level of groups of organisms, species, and clades? Can evolutionary patterns be caused by the forces of group, species, or clade selection?

The force of selection for traits of some class of entities can operate only when these entities are homogeneous with respect to these features and undergo processes analogous to birth and death of individual organisms. At first glance, it seems that groups of organisms as well as species and clades can meet these conditions. A large population of conspecific organisms may be subdivided into more or less isolated groups, and it is fully conceivable that such groups acquire some characteristic features of their own. Species of course differ in at least some traits—otherwise they would not be considered species—and their origination and extinction can be viewed as analogous to individual birth and death.

These two conditions, however, are necessary but not sufficient for groups, species, and clades to undergo selection of due to selection for their traits. Our flocks of birds could multiply due to the advantages the new behavioral feature confers upon individual organisms. In other words, the selection of such flocks could simply be an effect of natural selection at the individual level, with the feature spreading to ever new flocks because it provides migrants with a competitive edge while entering new habitats or invading other flocks. Our species of wingless grasshoppers could become more and more numerous simply because winglessness is, for ecological reasons, advantageous in the environment inhabited by these animals. Tetrapod clades could have increased in proportion within the vertebrates simply because they were encountering an extremely wide array of ecological opportunities for diversification, much wider than the ecospace available for fish diversification. In order to show that it is indeed the force of group, species, or clade selection that has brought about an observed pattern of selection of bird flocks, grasshopper species, or vertebrate clades, however, one has to know for sure that the pattern could not result from the action of selection at the individual level.

In the case of group selection, the considered phenotypic features must therefore be disadavantageous, if only slightly, for the individual organisms that have them. Adam Łomnicki analyzed such situations (Łomnicki and Hoffman 1987). The classic concept of group selection refers to the maintenance of individually disadvantageous traits in a large population consisting of several smaller, largely isolated groups that persist for a number of generations and are separated from one another by strips of inhospitable area. Łomnicki shows that group selection for a feature disadvantageous to individuals may successfully counteract individual selection solely when the groups have a considerable probability of extinction and the rate of migration from one habitat to another is relatively small. Even under such restrictive conditions, however, group selection is a very weak evolutionary force, capable of maintaining a trait in the population but not of displacing a trait that would be favored by individual selection. Another concept of group selection was developed by David Wilson (1980) who analyzed populations subdivided into groups of organisms that exist only temporarily but dissolve within the lifetime of each generation. In such a situation, it may be beneficial for an individual to behave altruistically toward its companions in a small group, because such behavior may increase the probability of survival and successful reproduction. Łomnicki shows, however, that although this evolutionary force may sometimes overwhelm the force of individual selection, but only if the individual disadvantage brought about by the altruistic behavior is very slight and if the groups are very small, including a handful of individuals and at best very closely related to each other.

Group selection thus is a weak evolutionary force that may poten-

tially operate under rather special conditions. In both situations discussed by Łomnicki, however, its power in any particular case is
ultimately determined by the effects that various individual phenotypic characters—whether they can be metaphorically described as
selfish or altruistic—have on the individual fitness. On the other
hand, the groups of individual organisms analyzed by Łomnicki and
other theoreticians interested in group selection can hardly be conceived of being a class of units of selection, simply because migration
of individuals from one group to another and the consequent in-
tergroup matings make such groups only loose associations that are
always vulnerable to invasion from outside. Traits that could possibly
be spread by group selection for them are bound to show, sooner or
later, a variation not only among but also within groups. Group selection, therefore, does not really operate at the level of groups of
individuals within a population or species, as would be necessary to
warrant its perception as a building block of the new, hierarchical
paradigm of evolutionary biology.

Species selection, by contrast, can in principle present a serious
challenge to the neodarwinian paradigm because it is supposed to act
upon entities fully separated from one another and each having a
number of features that vary among but are invariant within species
and that can be transmitted in the process of speciation to all daughter species. If such a character, invariant within a species and transmitted to its daughter species, were to increase the likelihood of
speciation or decrease the likelihood of extinction of the species as
compared to its related species having another character instead, it
might be maintained and spread within the clade or even the biosphere. Such a process of evolution of a clade or biosphere composition in terms of various categories of species could then be viewed as
driven by the force of species selection. Thus, the species appears not
only as benchmark but also as a potential unit of selection because
various categories of species can be subject to selection of due to
selection for their particular traits. Species selection might then be—
at least potentially—an evolutionary force that acts in addition to the
genetical forces and hence does not fit readily into the neodarwinian
framework. It would have to be regarded as a macroevolutionary
force whose operation in nature must be subject to specifically macroevolutionary laws.

That a force of this sort can conceivably operate in nature, does
not prove, however, that it really does operate. One could easily

conceive, for example, of a variety of parapsychological forces and even create a set of corresponding laws determining their action. But there may well be no parapsychological force at all. It is enough to have a look at the more ingenious science fiction books to realize how many forces and laws of nature can still be thought of. Stories have been concocted, for instance, about some special "bioenergetical" forces emerging from interaction among human brain cells; these forces are described as negligible, and in fact undetectable, until the brain reaches a certain critical mass, say, five or ten kilograms. This is not at all unthinkable, and it might even be empirically testable in the remote future when neurosurgery will advance so far as to be capable of performing the horrible operation of fusing together a number of different brains. Some strange psychological, neurological, or bio-chemical phenomena might then indeed be discovered. Perhaps a bioenergetical force of one or another kind might then be validly referred to as a, or even the, explanation. But until such phenomena are observed, the alleged bioenergetical forces are not needed to explain rationally the world around us. Even before testing whether a particular hypothesis should be accepted as explanation for some phenomenon, the phenomenon must be discovered that calls for such an explanation in the first place.

9.3 Evolutionary Trends as Macroevolutionary Phenomena

Species can be units of selection. The macroevolutionary force of spe-cies selection can operate in nature. It remains to be seen, however, if there are any historical biological phenomena that should actually be explained by the action of species selection. The claim that this is the case has become the core assertion of modern paleobiologists who advocate macroevolution as being causally decoupled from microevo-lution and who undertake to discover and describe some macroevolu-tionary laws, without which the course of evolution could not be prop-erly understood.

As a matter of fact, Hugo De Vries (1905) and Thomas Morgan (1903) wrote of "selection between species" as the mechanism of evolution primarily because they did not believe the force of natural selection at the individual level could possibly accomplish anything of significance in evolution. They envisaged selection between species as the main evolutionary force, which operates through differential survival of incipient species—originating chiefly by mutations—and thus decisively shapes the biosphere. This idea had been explicitly antidarwinian and, although it has not been entirely forgotten, it has been largely neglected, and abandoned by Morgan himself, ever since the actual power of natural selection was experimentally demonstrated.

In modern evolutionary biology, species selection was first invoked as an explanation for long-term evolutionary trends. Both Steven Stanley (1975), who coined the term, and Niles Eldredge and Stephen Jay Gould (1972), who had earlier introduced but not named the concept, were puzzled by the undisputable incidence, and even commonness, of very long sequences of fossil species that appeared to display a directional change in one or more phenotypic characters. The sequence of fossil horses, with their increasing body size, changing shape of the molar teeth, and transforming anatomy of the limbs, is a classic example of such a trend extending over millions and millions of years. Trends of this sort are what prompted the majority of paleontologists in the late nineteenth and early twentieth century to accept orthogenesis rather than selection as the main mechanism of evolution, for those workers simply could not reconcile the Darwinian claim for opportunism of evolution through natural selection, as they understood it, with the paleontological evidence for directionality of evolutionary change. The neodarwinian George Gaylord Simpson (1953) explained such long and directional sequences by assuming a persistent pressure of the force of natural selection, which he regarded as maintaining a trend as long as it progressed in a given direction—toward larger body size, for instance, or toward a modified shape of molar teeth in order to allow for more efficient grinding of the food—and then terminating it or changing its direction. To the advocates of punctuated equilibrium, however, the trend must appear as a paradox; for if there is no change within species, then how does the directionality of a sequence originate?

Given the strong version of punctuated equilibrium, which rules out any evolutionary change within established species, and even under the moderate version, which maintains that the evolutionary history of each species is largely stasis, Simpson's solution is obviously unacceptable, because evolutionary trends could not possibly represent patterns arising by continual, gradual, and unidirectional evolution of a sequence of species. Even assuming that evolution is gradual and unidirectional, however, does not solve the dilemma.

The time spans covered by some trends are so enormously long that the rate of phenotypic evolution within a trend appears very low, orders of magnitude lower than the rates observed in selection experiments of population geneticists or among living organisms. The selection pressures needed to account for this extremely low rate of evolution are so small that it becomes hard to believe they could persist unchanged for the time spans entailing an evolutionary trend. Russell Lande (1976) once calculated that a few selective deaths per million years per generation could suffice to account for the change of molar teeth during the entire evolutionary history of horses. This rate is so amazingly low that natural selection appears, paradoxically, too powerful an evolutionary force to explain the pattern of molar teeth evolution in horses, or in fact any other trend of this kind.

Confronted with this puzzle of evolutionary trends, Eldredge, Gould, and Stanley resorted then to species selection as the macroevolutionary force responsible for their origination. It is Simpson (1953) who emphasized very strongly that the evolutionary history of horses actually encompasses not only a single lineage, evolving from a small, four-toed animal in the Eocene to the modern horse, and not even a couple of parallel lineages, but a great many lineages, each evolving in a different adaptive direction. The so-called general trend in horse evolution is only a construct of human mind, which rather arbitrarily establishes a trend by simply comparing the anatomies of the earliest and the latest members of this phylogenetic bush and then filling the gap with other species whose anatomies, when taken together, constitute a consistent series. According to Simpson, the very possibility of constructing such a trend implies, however, that in spite of the wide variety of selective pressures which forced various horse lineages to evolve in different directions, one may also speak of a persistent pressure of natural selection forcing this one particular lineage to evolve continually in the same adaptive direction.

The proponents of punctuated equilibrium reject this explanation because it is incompatible with what they claim to see in the fossil record—that is, stasis punctuated solely by rare events of speciation—but they accept Simpson's view of what constitutes an evolutionary trend. They also assume that the phenotypic differences between all daughter species and their parent species in the phylogenetic bush are randomly distributed relative to the direction of a long-term trend, no matter how the trend is constructed. The latter assumption makes an explicit analogy between speciations and mutations, both these categories of events introducing variation at random relative to evolution. It has been christened Wright's Rule by Gould and Eldredge (1977), in honor of the great neodarwinian Sewall Wright who thought of speciation in this way. The conjunction of the strong, or even the moderate, version of punctuated equilibrium with Wright's Rule leads then to the concept of species selection as the explanation for evolutionary trends. Trends are supposed to be due to an increased speciation rate of a certain kind of species (for example, horse species with larger body size and reduced toes, barnacles with solid wall structure and low growth rate as contrasted with their relatives with porous structure and faster growth, or fully infaunal, mobile bivalves actively burrowing in bottom sediments as opposed to sessile brachiopods and bivalves permanently attached to the substrate) or to a decreased extinction rate of such species, or to their greater duration while the rate of speciation is approximately constant among the considered groups or kinds of species. Trends thus represent changes in frequency distribution of various kinds of species. They amount to patterns arising from selection of one or another category of species within a clade, an ecological realm, or even the whole biosphere. Given this interpretation of evolutionary trends, species selection appears as a potential explanation.

Not all evolutionary trends, however, must necessarily be of this sort. It is perfectly plausible that a given trend reconstructed from the fossil record by paleontologists represents only a single lineage of species linked by ancestor–descendant relationships, rather than a branching phylogenetic bush from which the paleontologist arbitrarily chooses a unidirectional sequence. An evolutionary trend may not be equivalent to a pattern of change in the relative abundance of species within a larger taxonomic or ecologic category. Quite a number of case studies from the fossil record could be cited as examples of such trends including only a series of ancestors and descendants. For instance, Adam Urbanek (1963) describes from Upper Silurian strata a lineage including four successive linograptid species that constitute a trend from a typical graptolite colony with individuals located all along a single stipe to a complex, multibranched colony. Another long-term evolutionary trend in a single lineage of graptolites is described by Cooper (1973); examples seem to abound also among conodonts and ostracodes. It is of course impossible to rule out the hypothesis that these trends also are isolated from a phylogenetic bush. There is no way to test and refute the hypothesis that other related lineages of these animals were simultaneously evolving in directions that have not yet been discovered by paleontologists, or they were so short-lived that the chances of their preservation in the fossil record have been reduced to almost none. The only assurance the paleontologist can find in such a situation is in the overall quality of the record of these animals, which appears to be very good for graptolites, conodonts, and ostracodes. In any event, however, such trends are not constructed arbitrarily by the mind of the paleontologist; naturally, the historical biological phenomenon one would like to explain is what one actually finds in the record rather than what one can imagine to have existed.

The term "evolutionary trend" evidently refers to two clearly distinct kinds of historical biological phenomena. On the one hand, it is often used to describe a pattern of morphological transformation in a sequence of species over a very long time span; the linograptid graptolites analyzed by Urbanek provide a good example. On the other hand, it is often employed as the label for patterns of a relative change in number among various groups of species or even higher taxa. The difference between these two meanings of the term reflects two divergent approaches to the history of life on the Earth, two contrasting foci of interest to historical biologists. Niles Eldredge (1979) introduced the concepts of "transformational" and "taxic" approaches to evolution to denote this very distinction.

Eldredge originally thought of these two approaches as stemming from the conceptual opposition between phyletic gradualism and punctuated equilibrium, because under the strong version of punctuated equilibrium morphological transformations in evolution are negligible while the relationships among species as cohesive entities and among various kinds of species are what constitutes macroevolution. There is, however, no logical link between taxic approach—as I understand it, in contradistinction to transformational approach—and punctuated equilibrium. To take the taxic approach to historical biology simply means to consider species, or even higher taxa, as benchmarks of evolution; it reflects an interest in patterns of change in taxonomic structure of the biosphere rather than in the ways particular species and their more or less exquisite and more or less clumsy adaptations to environment originated and evolved. There has been a long and highly respected tradition of taxic thinking in paleontology, personified recently by Thomas Schopf and Arthur Boucot, two of the most vociferous opponents to punctuated equilibrium. In fact, evolutionary trends of taxic type—on any temporal and taxonomic scale—can be recognized and described regardless of whether they encompass either punctuated, or gradual transformational trends at the species level.

9.4   Transformational Trends

Within the framework of the transformational approach, trends are not patterns reflecting selection of certain kinds of species—no matter whether or not one subscribes to the punctuational view of evolution. They cannot therefore be explained by species selection. In the strong or the moderate version of punctuated equilibrium, a temporal sequence of fossils that constitute a unidirectional morphological series can well be explained, however, by a succession of speciation events that were taking each descendant species consistently in the same adaptive direction relative to its ancestor. Even in the absence of any significant evolutionary change within the lifetime of any particular species in a sequence, such directed speciation, as Steven Stanley (1979) calls it, produces a transformational trend.

Directed speciation is obviously at odds with Wright's Rule. If the latter were true, purely transformational trends would be impossible. Then, given the strong version of punctuated equilibrium, the force of species selection would indeed provide the only conceivable explanation for evolutionary trends. Wright's Rule, however, has never been tested, let alone corroborated, in the fossil record, so it is not known whether the condition that may make a reference to species selection necessary is actually met. Directed speciation, in turn, is quite plausible on theoretical grounds. Speciation is most probably opportunistic instead of strictly random. The point is that even though many marginal populations are adapted to habitats within a wide spectrum of environmental conditions atypical for the species, and even though they consequently differ in various phenotypic aspects from the mainstream populations of the species, not all of them are incipient species. In order to survive and evolve in isolation long enough to establish themselves as new species, such populations need to encounter an ecological opportunity. Such opportunities often occur along environmental gradients—climatic, for example, as encountered while moving northward along the Pacific Coast of North America, or toward a mountaintop in the Rocky Mountains—and, consequently, the phenotypic adaptations selected for in new species located along such a gradient may also be arranged in a consistent series, because adequate adaptive response to more of the same environmental challenge may often involve more of the same feature; for instance, increasing body size under cooler climatic conditions or more specialized feeding mechanisms under scarcer food resources. Therefore, the availability of ecological opportunities for evolution of new species can in fact account for the many cases of apparently directed speciation along environmental gradients described, for instance, by Verne Grant (1963) and Ledyard Stebbins (1982) among living plants.

Directed speciation can also be promoted by a tendency among many species to meet some environmental challenges by shifting the timing of somatic growth and reproductive maturation relative to each other. Depending on the environmental conditions, it may be advantageous for the organism's reproductive success either to mature and reproduce as early as possible or to postpone reproduction until rather late in life, that is, to achieve maturity either at a juvenile or a fully adult or even gerontic ontogenetic stage. Given such conditions, therefore, there will be selection for either acceleration or delay of sexual maturation relative to somatic development. Such selection may eventually lead to the origin of new species, differing in one way or the other from its ancestral form. This is the phenomenon of heterochrony in evolution. A classic example is the Mexican salamander, which maturates while retaining its juvenile morphology, with gills as the most characteristic feature; an equally classic but contrasting example is the gigantic Irish Elk from the Pleistocene, which evidently grew to much larger body size than any of its relatives before attaining maturity. Along an ecological gradient, then, a sequence of species with progressively accelerated or delayed maturation, and hence with progressively juvenile or gerontic anatomy, can appear in the course of evolution. The successive speciation events will be directed and form an apparent evolutionary trend of the transformational kind. Kenneth McNamara (1982) documents in fact a number of such trends among fossil brachiopods, trilobites, and echinoids, each of which seems to be best interpreted as reflecting a sequence of speciation events through heterochrony leading toward either earlier or delayed maturation.

Transformational trends, however, do not have to follow the pattern of punctuated equilibrium. And if they do not conform to this pattern, then directed speciation appears no longer the main, or at least the most plausible, mechanism responsible for their occurrence.

The dilemma that natural selection at the individual level is too power-ful an evolutionary force to account for the very low rate of pheno-typic, and presumably also genetic, evolution in many trends of this sort can also have an even simpler solution.

There is no reason to suppose that evolution should proceed for long at a constant rate and in the same direction. On the contrary, there are good reasons to expect the rate and direction of evolution to vary along a lineage in response to changing environmental condi-tions and changing population size of the species. This expectation is in fact embodied in the weak version of punctuated equilibrium as well as in all reasonable versions of phyletic gradualism. The ex-tremely low rates calculated by Lande and others, however, are only the average figures over very long time intervals, that is, millions and millions of years. During such a long time, the rate of phenotypic evolution may be repeatedly accelerated and slowed down; evolution may go for a certain time in a direction different from the trend, and then the previous direction may be resumed again, thus giving the false impression of stasis within a trend in evolution of a given fea-ture. For instance, a trend in teeth evolution of a carnivorous mam-mal may well include a period of apparent stasis, when the teeth essentially did not evolve at all, while improvements in the locomo-tory mechanism and in the physiology of feeding were the prime mechanisms of increasing individual fitness. Thus, the overall rate of teeth evolution may be low, but it would simply be an artifact of the averaging procedure, whereas the actual rates would be higher and could be explained by stronger and more variable selective pressures as the stimuli for evolution.

Richard Dawkins (1982) makes an instructive analogy. If one throws a cork into the ocean in, say, Florida, it may take it much longer to travel across the Atlantic than it could be expected on the basis of the average velocity of the Gulf Stream. This cannot be taken to imply, however, that the cork does not travel with oceanic cur-rents. Local currents and strong waves, caused occasionally by stormy winds that it would happen to encounter on the way, may push it sideways and even backward on its route. There is no reason to expect it to float unidirectionally. Still, the moving force of its travel would remain the movement of water. Similarly, unidirectional evolu-tion at a constant rate is not the only, nor even the most likely, alternative to complete evolutionary stasis. And if another alternative happens to be true, evolutionary trends of transformational kind become a macroevolutionary phenomenon that is perfectly understandable within the neodarwinian framework of microevolutionary mechanisms of evolution.

9.5   Taxic Trends

The situation is very much the same with many evolutionary trends of taxic type, even though they involve changes in frequency distribution of various kinds of species. Taxic trends represent patterns resulting from selection of species and might be potentially explained by selection for a certain set of specific characters. They nevertheless fit the neodarwinian paradigm with its emphasis on individual selection. In the majority of instances discussed by Stanley and his colleagues under the heading of species selection (for example, Stanley 1979, Stanley and Newman 1980, Stanley et al. 1983), the increase of one group of species relative to another one is explained by the differential responses of representatives of these groups to a variety of selective pressures, such as interspecific competition, predation, or physical habitat alteration. For instance, sessile and rather thin-shelled benthic invertebrates permanently attached to the substrate, which had dominated marine shallow-water habitats in the Paleozoic, gradually declined in species number in the Late Mesozoic and Cenozoic, giving way to highly active, rapid burrowers living within bottom sediments and to various forms with strongly ornamented and very thick shells. This trend is paralleled by an increasing trend of predatory, durophagous crabs and fishes capable of crushing the invertebrate shell or carapace. Geerat Vermeij (1977) and Steven Stanley (1979) explain this trend in benthic invertebrate prey species as being driven by the force of species selection for morphological and behavioral traits advantageous for resistance to durophagous predators. Predation must have been heavier on weakly armored prey species, and these differential predation rates certainly led to higher extinction rates among species that were more strongly fed upon, and perhaps also to their lower speciation rates through extirpation of many small populations that could otherwise establish themselves as new species.

This explanation is sound and rather uncontroversial—that is, except for the reference to species selection in this context. The pattern observed by Vermeij and Stanley can also be a strict, fully predictable consequence of the action of natural selection at the individual level alone. Predation by durophagous crabs and fishes must have exerted a very strong selective pressure for adequate morphological and behavioral responses among invertebrate prey species. Among bivalves and gastropods, the selection for thicker and more strongly ornamented shells and for more efficient locomotion caused selection of individuals equipped with genes that could bring about those traits within each particular species, and consequently these predator-resistant features spread in the marine realm at the expense of the previously widespread sessility and thin-shellness. This explanation in fact only rephrases Vermeij and Stanley, but without invoking species selection. Here the observed taxic trend, or the pattern of selection of certain species, is explained entirely in terms of individual selection, as it could also be done by Richard Dawkins within the framework of his selfish-gene concept.

More generally, natural selection at the individual level must lead to evolutionary trends of this kind, because there are evolutionary agents that affect in a similar way quite a number of independent lineages or groups of species. On the other hand, interspecific interactions and autecological responses are not characteristic of species as opposed to individuals. To the contrary, there is a variation in these aspects within each particular species. It is individual organisms, not species as entities, that compete, feed upon one another, and die due to, say, a rapid drop in humidity or salinity of the habitat. Therefore, such evolutionary trends as an increase or decrease in number of a certain category of species due to predator or parasite pressure, competition, or habitat alteration are not phenomena that would demand explanation by species selection. They are not patterns of selection of species due to selection for their species-level characters, but patterns of selection of species expressing a selection of individuals caused by selection for their individual attributes.

If the mechanism responsible for an increase in number of species equipped with a certain trait is ultimately the adaptive advantage this trait confers upon individual organisms, there is no reason to regard species selection as the driving evolutionary force. It does not follow, however, that where no selection for a character occurs at the individual level, the character's spread in a taxon or the biosphere through selection of species that carry it is indeed due to selection for such species. In other words, even if selection of species is not caused by individual selection, it still may not be caused by selection for species properties.

Evolutionary trends of taxic type may well result from historical accidents causing some categories of species to decline or even disappear, whereas other kinds of species take over due to sheer good luck rather than due to selection for any of their characteristics. For example, if the Krakatoa island had a rich ant fauna including an endemic genus whose closest relatives also lived on other islands in Indonesia, the volcanic eruption and total destruction of the Krakatoa island would not only exterminate all ant individuals living on the island but it would also effect a change in species composition of the ant fauna in the entire Indonesia. There would be a selection of the more cosmopolitan ant taxa without selection for any of their particular properties.

Such a pattern does not have to arise by catastrophic extinction of one or another group of species. It may also originate gradually, and hence be more rightly called an evolutionary trend. A good example is the evolution and then disappearance of the rich fauna of endemic cockles in the Eastern Paratethys over millions of years of the late Tertiary. Paratethys was an extensive marginal sea developed in the late Tertiary north of the mountain chains including the Alps, the Carpathians, and the mountains of Turkey and Iran. During much of the time it existed, it was linked through more or less wide straits to the Mediterranean area. In the latest Miocene and Pliocene, however, its eastern part became isolated and finally disappeared. During its separate existence, this Eastern Paratethys provided opportunities for the famous Pontian radiation of cockles, which of course later became extinct. The pattern of first increase, and then decrease down to zero, of the proportion of the species of these Pontian cockles within the entire family of cardiid bivalves clearly is a macroevolutionary pattern of selection of a certain kind of species. It is caused, however, by the historical accident that ultimately cut off the Eastern Paratethys from the world ocean rather than by selection for any features of either the Pontian, or the other cockles.

Generally, evolutionary trends of taxic type often are a macroevolutionary phenomenon caused by selection of species without selection for their particular traits. Elisabeth Vrba's (1980, 1983) effect hypothesis, or effect macroevolution, puts forth this explanation for trends. The word "effect" is employed to emphasize that the resulting macroevolutionary pattern—a taxic evolutionary trend—is merely a byproduct, a side effect of quite ordinary evolutionary forces operating within populations and species. When Vrba introduced her hypothesis, she actually wrote of both taxic and transformational trends simultaneously. She discussed the morphological change in antler shape that occurred since the Miocene among bovid antelopes in Africa, and also the pattern of change in numerical proportions of the two subgroups of these antelopes, with one subgroup becoming with time much more speciose than the other.

In the case of transformational trend, Vrba's explanation simply implies that a substantial phenotypic change takes place in a lineage without there being a selection for the new phenotypic variants. It thus refers to genetic drift, which is capable of fixing not only a neutral but even a slightly deleterious gene in a population, or to gene hitch-hiking; the latter term denotes the maintenance and even spread of a gene due to its obligatory linkage to another gene that is adaptively advantageous to its carrier at the individual level and is therefore selected for. There is nothing special or implausible about genetic drift and gene hitch-hiking as explanations for some phenotypic patterns in evolution, but I doubt if they can indeed provide an adequate explanation for long-term evolutionary trends. That is, unless the trend in question involves no more than a change in such a simple phenotypic trait as the color of eyes in humans.

It strikes me as rather improbable that more complex phenotypic traits—for instance, the structure of the vertebrate eye or even the shape of the antlers in bovid antelopes—could have so little impact on individual fitness that they would evolve for long in the same direction without much interference from selection. Most likely, phenotypic features of this sort are continually under pressure of natural selection due to their continuing contribution to individual fitness. They consequently acquire a complex polygenic background which makes their actual phenotypic expression less vulnerable to the vagaries of historical accident during individual development. Max Hecht and I have in fact argued in our recent critique of the modern macroevolutionary challenges to neodarwinism (Hecht and Hoffman 1986) that such complex traits are maintained, let alone evolve, primarily by the force of natural selection, and that once natural selection ceases to act to this effect, the polygenic background of a trait will disintegrate due to mutation and recombination and the trait itself will degenerate rather than be transmitted along the evolving lineage.

The history of the eye loss among cave fishes can serve as an instructive example. A Mexican freshwater fish, *Astyanax mexicanus,* has fully developed eyes. Its descendants confined to deep cave systems, however, have no use for eyes in their permanently dark habitat. Evolution has in fact led to the loss of the eye in all those underground populations and species, but in a completely disorderly fashion, with different structures being reduced to a different degree in various populations (Cahn 1958, Wilkins 1971). This example of what happens to a complex and functionally efficient phenotypic structure once it ceases to be an adaptation for one or another biological function—that is, once it ceases to be under the control of natural selection—suggests that such traits evolve largely through natural selection, and not by genetic drift. Natural selection is furthermore capable of disrupting genetic and developmental linkages between various traits, so that gene hitchhiking does not seem to have much of a chance to overwhelm natural selection over very long periods of time. It is therefore unlikely that transformational trends involving any more complex genetic phenomena than single gene fixation by genetic drift could be explained as an accidental effect or byproduct.

The situation is very different for evolutionary trends of taxic type, however. The imaginary example of Krakatoa ants and the actual case of Pontian cockles both corroborate Vrba's assertion that substantial changes in frequency distribution of various categories of species within a larger taxon can be brought about by factors entirely accidental to the biological traits that might conceivably be selected for or against at the species level (that is, traits that might possibly contribute to either increase or decrease the average speciation rate, extinction rate, or duration of species in one category or another). By analogy to genetic drift and gene hitchhiking, John Maynard Smith (1983) and John Turner (1986) introduced the terms "species drift" and "species hitchhiking" to denote the mechanisms of trend origination by such accidental factors. Species drift means that the macroevolutionary pattern of a taxic trend toward dominance of species equipped with a certain trait has arisen by chance alone. Species hitchhiking, in turn, means that such a trait has spread due to its accidental association with a biological characteristic that actually enhances the potential for speciation or long-term species persistence, or diminishes the vulnerability to species extinction.

I am unaware of any good real example of species hitchhiking but I think I may put up an imaginary one. Consider, for instance, two closely related phytophagous insect species in a tropical rain forest that differ in their pigmentation patterns. Because of unspecified ecological circumstances, one species goes for food specialization, whereas the other remains a food generalist and feeds on whatever leaves it finds. The former species then becomes much more likely to speciate than the latter; for example, due to speciation in the food plant. Provided that, for some anatomical and physiological reasons, the food specialist will tend to beget food specialist species rather than generalists; and provided also that the ecosystem is sufficiently stable and resourceful to allow for survival of all, or at least the majority of, descendant species, the original specialist species will give rise to much more numerous a group of species than its more generalized relative will. Therefore, the pigmentation pattern of the specialist will also show up in much more speciose a group. This spread of its pigmentation pattern could be a case of species hitchhiking, and there is little doubt that such phenomena must be quite common in nature.

Species drift has most certainly often occurred in the history of the biosphere, and it may even be among the main mechanisms of evolution as viewed from the taxic perspective. Every extinction event, and especially every case of a wholesale community extinction, due to physical habitat alteration or its destruction leads to a more or less substantial change of the taxonomic composition of the biota. To the extent that the physical environmental change involves factors different from those responsible for variation in fitness within particular species—for instance, desiccation of a lake system for freshwater fish—the effects of extinction are to be explained by species drift. Similarly, every case of enhanced speciation in a taxon due to its accidental encountering of environmental opportunities—for example, tectonic fragmentation of a land mass into several smaller islands, which provides opportunities for geographic isolation and differential selective pressures—also is an example of species drift.

Thus, the effects of species drift are far from trivial, and the process indeed has not been previously named, let alone emphasized, within the framework of the neodarwinian paradigm of evolution. What the occurrence and even importance of species drift ultimately means, however, is only that evolution always takes place in an environmental context, and that changes in this context are as important to evolution as are changes in the biological composition of species.

There is no macroevolutionary law here, though; no macroevolution-ary process operates upon species as entities of their own.

9.6 Species Selection as Causal Explanation

This is not to say, however, that evolutionary trends of taxic type can never be explained as macroevolutionary patterns arising by true species selection, that is, through selection of species caused by selec-tion for their specific properties. Max Hecht and I have outlined three basic conditions that must be met for genuine species selection to be invoked to explain a taxic evolutionary trend (Hecht and Hoffman 1986).

First, the trend must represent a pattern calling for explanation that could, at least in principle, be phrased in terms of species selection. It must be a pattern of two closely related and monophyletic taxa, or clades—each taxon stemming from only one original species—that differ in species number at a certain point in their parallel evolutionary histories; the very fact that each taxon started with only one species would then ensure that a change in frequency distribution of their representative species within the higher-rank taxon has indeed taken place through time. It is important that the compared taxa have equally long evolutionary histories, so that the factor of time can be neglected; if not, then the less speciose taxon should be of older geological age. Ideally, they should be sister taxa, originated by a single speciation event that split an evolving lineage and thus gave rise to two closely related clades. The two compared taxa must furthermore differ in some characteristics that show no variation within particular species and are inherited from the respective original species. It is only with respect to such species-wide traits characteristic of entire clades that one can speak of taxic trends. If the compared taxa are not monophyle-tic, however, if they are polyphyletic—that is, artificial groupings of species brought together by the systematist because of their pheno-typic similarity but in spite of their nonrelatedness, or different genealogies—there is no reason to suppose that a single biological process is responsible for the observed difference in their speciosity. One is then comparing apples with oranges, or better—the diversity of apples with the diversity of oranges and all other ball-like objects.

Given the notorious limitations of the fossil record, this condition can never be demonstrably met in paleontology. For even if a case for the monophyletic nature of some taxa can be made and strongly corroborated, an actual difference in species number between taxa cannot be proven. On the one hand, cryptic or sibling species may occur much more often in one of the compared taxa than in the other; and, being indistinguishable in the fossil record, they may distort the picture of differential speciosity. On the other hand, the vagaries of fossilization and later preservation may also substantially change the actual biological pattern without leaving many clues as to what has been preferentially removed from the record. Macroevolutionary patterns that meet our condition of monophyly and difference in speciosity can nevertheless be found in the living biosphere—for example, the difference in speciosity between the articulate and the inarticulate brachiopods may be of this kind—and even in the fossil record it would probably be too much of scientific purism to reject such possible instances solely because they might perhaps be biased. Provided that the criteria of monophyly, approximately equal geological age, and a substantial difference in fossil species number are met, a taxic trend established by paleontologists can, I think, also be regarded as calling for explanation by species selection.

The second condition is that at least one of the species-wide characteristics that make up the biological difference between two compared monophyletic taxa—and therefore either spread or decline through time in terms of the number of species that carry them—can indeed be supposed to cause the observed pattern of difference in speciosity between these taxa. A biological property of one taxon must be identified that contributes to either increase the potential for speciation, or to decrease the vulnerability to species extinction, as compared to representatives of the other taxon. In other words, a species trait must be identified that is selected for and thus causes a selection of species that bear it. If this condition is not met, we are back again to species drift, to effects that historical accidents have on the processes of evolution.

The third condition to be met by any species-selectionist explanation of a taxic trend is that the inferred macroevolutionary force of selection for a certain species-wide characteristic, or a set of characteristics, must be irreducible to the force of individual selection. Its action cannot possibly be replaced by natural selection at the individual level. This postulate is not uncontroversial. It is perfectly reasonable that both these forces operate simultaneously and jointly shape a macroevolutionary pattern. Egbert Leigh (1977) developed an argument that those taxa should best proliferate in evolution where natural selection within populations works more nearly for traits that could also be favored by species selection. The snag is, however, that it would be terribly difficult to demonstrate under such conditions that species selection does indeed operate in any particular case. It would be difficult to show that it is a real evolutionary force. For example, the increase in speciosity of well-armored relative to thin-shelled benthic marine gastropods in the late Mesozoic and Cenozoic may well be partly due to genuine species selection for resistance to durophagous predatory crabs and fishes, because such resistance is likely to decrease species vulnerability to extinction; but under conditions of increased predation pressure, individual selection for predation-resistant features must certainly lead to a spread of these traits in a wide variety of species, and hence the actual, causal efficiency of species selection cannot be directly demonstrated. In such a situation, there is no need to invoke species selection, since natural selection at the individual level provides an adequate explanation for the observed evolutionary trend.

Thus far, not a single empirical example of purported species selection has been described that meets all three conditions, although a number of macroevolutionary patterns have been repeatedly cited in this context. For instance, Thor Hansen (1980, 1982) documents for a number of Tertiary marine gastropod families in southeastern North America a considerable decrease in proportion of species with planktonic larvae capable of active feeding at that developmental stage, and hence of spending much time as plankton and of dispersing over very long distances in the ocean. Species with such planktotrophic larvae, however, generally have longer durations in geological time than their relatives with nonplanktotrophic larvae that feed upon yolk and rapidly settle on the sea bottom. Hansen suggests that the pattern of change in frequency distribution of species with planktotrophic versus nonplanktotrophic larvae is due to the greater potential of the latter for speciation. Apparently, the relatively low dispersal rate of marine gastropods with nonplanktotrophic larvae enhances isolation of their local populations and leads to increased speciation. This gastropod example thus seems to meet the second condition on our list—there may exist a species-wide trait, the nonplanktotrophic mode of larval development, that contributes to increased speciation and thus might be selected for at the species level.

It does not meet, however, the first and the third conditions listed above. The phylogeny for the gastropod taxa compared by Hansen is largely unknown, and it is debatable whether the groups with planktotrophic and nonplanktotrophic larvae are monophyletic. In fact some data reported by Hansen indicate that, among the marine gastropods he considers, species with nonplanktotrophic larvae arose independently in several lineages that had originally had larvae capable of feeding while floating as plankton in the ocean. If so, however, it is only natural that the group of species with nonplanktrotrophic larvae is more numerous than the other. Moreover, the amount of species origination by transformation of an earlier species, without true speciation, and the amount of immigration of species with nonplanktotrophic larvae from outside the study area are unknown, although there are some indications in Hansen's data that these factors may not be negligible. The empirical pattern observed by Hansen does not call for explanation in terms of species selection. The pattern of independent appearance of nonplanktotrophic larval development in various gastropod families also suggests a conclusion that does not appear to be very surprising: that the mode of larval development is under a strong control by natural selection at the individual level. For some ecological reasons, the nonplanktotrophic larval development turned out to be adaptively advantageous for these gastropods in the particular environmental setting investigated by Hansen. No direct clues are left in the record that could suggest what particular environmental challenges were most easily and adequately met by this mode of larval development; though one might for example conjecture that increased environmental stability and predictability after a major marine transgression could have played a role. The important point is that given the evident feasibility of evolutionary transition from planktotrophy to nonplanktotrophy of the larvae, that is, given the absence of developmental or physiological constraints that would make such evolution impossible or at least highly improbable, there may be no need to invoke species selection as explanation for the observed pattern—even if the pattern would eventually turn out to be indeed a case of closely related monophyletic groups that change their speciosity relative to each other in evolution.

David Jablonski (1986c) takes much further the case for the modes of larval development in marine gastropods being subject to species selection. He provides compelling quantitative data to show that monophyletic groups of species with nonplanktotrophic larvae do indeed have greater speciation rates than those with planktotrophic larval development, and he suggests that the mode of larval development is highly heritable at the species level—that is, transmitted during speciation events—because evolutionary changes from planktotrophic to nonplanktotrophic development seem to be nonexistent in the late Cretaceous gastropods he studies. This is about as much as one could demand. The conditions are met for the force of species selection to bring about a clearcut taxic evolutionary trend for larval nonplanktotrophy. Moreover, such a trend can be expected to appear also in other groups of benthic marine organisms (e.g., bivalves, crustaceans), since the planktotrophic versus nonplanktotrophic modes of larval development are very broad categories, unrestricted to the class Gastropoda. And yet there is no universal trend of this kind in nature, not even among the late Cretaceous gastropods studied by Jablonski. What is then the empirical pattern that species selection for nonplanktotrophy should explain? Why is this alleged macroevolutionary force inefficient in this case?

At least three different solutions to this apparent paradox can be given. First, the low dispersal potential of nonplanktotrophic larvae is likely to enhance not only high speciation rates but also high extinction rates. This is in fact documented by the association of high speciation rates with shorter durations among the gastropod species investigated by Hansen and Jablonski, and Steven Stanley (1986) provides good data of this kind for bivalves. On balance, therefore, there may be no species selection for either mode of larval development. Second, various environmental correlates of nonplanktotrophic larval development and its consequent low dispersal capabilities can be responsible for both increases in speciation and extinction in such organisms. Geographic fragmentation of otherwise fairly stable habitats may serve as an example. It leads to an association of frequent speciation with frequent extinction of species that generally inhabit small and isolated patches of environment. This is the fission effect recently described by Steven Stanley (1986). Under such conditions, which Stanley envisages for marine bivalves, there would be no species selection for one or the other mode of larval development, but only species drift induced by the environment. Third, the gastropod groups compared by Jablonski differ not only in their mode of larval development but also in several other biological traits; perhaps one of those other characteristics is what is really selected for at the species level. This seems to me the least likely solution because it is entirely ad hoc, without even the faintest evidence to support it, but the other two appear to be quite plausible and are in fact as well, or as poorly, supported by evidence as is the claim for the role of species selection itself.

Even less convincing are all the other examples given thus far by the proponents of species selection as an important macroevolutionary force. Steven Stanley (1979) argues that the greater speciosity of short-winged than long-winged forms in the grasshopper genus *Melanoplus* results from a lower capability of dispersal, and hence greater potential for speciation, in the former. This argument resembles Hansen's and Jablonski's, but its flaw is even more evident—on zoological evidence, short-winged species of *Melanoplus* surely do not constitute a monophyletic subgenus or species complex but have independently arisen in a number of lineages. Stanley also suggests that the enormous species diversity of various organisms in tropical rainforests of South America, as compared to their relatives in more arid areas, is to be explained by species selection. This is a clear case of species drift, however, because the mechanism of increasing the speciation rate in the rainforest birds and lizards mentioned by Stanley did not involve any biological property of the organisms but only a repeated fragmentation of the habitat during the Pleistocene climatic oscillations. Other alleged instances of species selection involve the apparent decrease in variety of shell shapes among marine gastropods over the geological time scale (Gilinsky 1981), the undisputable increase in the contribution that the Hawaiian fruit fly species make to the entire family of drosophilids (Gilinsky 1986), and many other macroevolutionary patterns. None of these historical biological phenomena, however, necessitates a reference to the macroevolutionary force of species selection as explanation (Hecht and Hoffman 1986).

Leigh Van Valen (1975) describes—though under the heading of group selection—selection of genera of small mammals and foraminifers relative to large ones. He also shows that this pattern is due to the rate of genus formation being higher for clades of small rather than large organisms in these groups. This might be a phenomenon of clade selection, but the reservations I have expressed about the plausibility of clade selection beyond the species level are fully applicable in this case. None of the compared groups of taxa (large and small mammals, large and small foraminifers) is monophyletic; though evolution only exceptionally proceeds toward size decrease in mammals and foraminifers, thus rendering negligible the failure of this pattern to meet the monophyly condition. Small size per se, however, is not a trait that could contribute to increase the rate of genus origination. It is only a correlate of many other biological and ecological characteristics. The pattern observed by Van Valen appears therefore to be caused by a combination of clade drift and hitchhiking rather than genuine clade selection.

Certainly the strongest case for species selection as explanation for an actual macroevolutionary pattern has been made by John Maynard Smith (1978, 1983). Maynard Smith notes, as did many others before him, that the widespread occurrence of sexual reproduction in the biosphere is a sort of paradox that calls for explanation. Every female transmits only a half of its genes to each offspring, whereas it would transfer the totality of the genome if it were to reproduce parthenogenetically, without any contribution from a male. This is the so-called genetic cost of sex. Given that parthenogenetic reproductive systems can develop from normal biparental sexual reproduction—and there are many examples to show that such a transition is possible in evolution—parthenogenetic females should therefore be favored by natural selection at the individual level. Yet the world is full of species reproducing sexually.

The traditional, neodarwinian solution to this paradox refers to the adaptive advantage that genetic recombination during sexual reproduction should bring about in heterogeneous, unstable, and unpredictable environments. In such an environment, even a perfect adaptedness of an organism to its current habitat does not guarantee that it would also be successful in the habitat of tomorrow. Therefore, it does not pay to have offspring identical, or even very closely similar, to oneself. In fact, genetic variation among progeny increases the likelihood that at least some of them will achieve maturity and successfully reproduce. Hence, sexual reproduction is promoted. Several mathematical models of natural selection suggest, however, that under realistic environmental conditions this individual adaptive advantage is too weak to overwhelm the genetic cost of sex. Maynard Smith argues that a part of the solution to the paradox of sex relates to species selection. Sexual reproduction is not only advantageous to the individual in variable environments but it also contributes to a decrease in species extinction rate. Parthenogenetic species are likely to be less variable genetically and therefore more vulnerable to extinction by unexpected vagaries of the environment. According to Maynard Smith, it may be the joint effect of individual selection for genetic variability of the offspring and of species selection for genetic variability of the entire species that outweighs the genetic cost of sex and causes the taxic trend expressed in selection of sexually reproducing species.

Even this case is not compelling, however. On the one hand, sex is not necessarily a species-wide characteristic. Many organisms, including vertebrates, are facultatively parthenogenetic, which indicates that some unknown but powerful evolutionary forces at the individual level counteract natural selection for decreasing the genetic cost of sex. On the other hand, Graham Bell and Maynard Smith (1987) have recently discovered that an additional and rather considerable adaptive advantage of genetic recombination appears at the individual level if the biotic environment of a species is taken into account. Complex biotic interactions are likely to promote genetic variability and hence sexual reproduction, and parthenogenetic species do indeed occur most commonly in relatively simple ecosystems rather than in complex ones. It is therefore very likely that a combination of selective pressures arising from environmental unpredictability and complex interspecific interactions may suffice to account for the commonness of sex in the biosphere.

What we are left with as the only unquestionable case of species selection is the thought experiment discussed by Anthony Arnold and Karl Fristrup (1982). If we take two species of grasshoppers that differ exclusively in that one species possesses a gene for speciation, while the other species does not, and if we keep them in such conditions that the number of individual organisms descendant from both the original populations is maintained constant, we shall get a difference in speciosity between two monophyletic groups. The former species will give rise to many more species than the latter, although there certainly will be no natural selection at the individual level to this effect. Many philosophers have accepted this argument as evidence for species selection. Biologically, however, this example is so utterly unrealistic that it cannot be regarded as evidence for the actual operation of selection at the species level.

It is inconceivable that two species would differ solely in presence or absence of a gene for speciation. There is no propensity for speciation that could possibly be considered in terms of a genetic, hereditary trait. Speciation rate is merely a net result of the interplay of all biological characteristics and the environmental change during the evolution of each lineage or higher taxon. It is not a biological trait of its own, intrinsically characteristic of any taxa. Consequently, however, the thought experiment brought up by Arnold and Fristrup shows only that species selection might be a causal force in evolution but not that it is one.

Thus, the status of species selection as a causal macroevolutionary force reaching beyond those envisaged by the neodarwinian paradigm of evolution is dubious. It is certainly conceivable that such a force does operate in nature. This was recognized long ago by Ronald Fisher (1930), Sewall Wright (1945), and George Williams (1966), although they did not spell out the problem very clearly and they certainly did not believe such a force may indeed be causally efficient in evolution. The unfortunate linkage between species selection and punctuated equilibrium considerably muddled the issue, for there is clearly no necessary logical connection between the two concepts. Arnold and Fristrup (1982) did a great job in helping to clarify the situation and to convince everybody that there may exist patterns of evolution to be explained by species selection. Species selection is by now firmly established as a potential evolutionary force. And if it really operates in nature, then the neodarwinian paradigm of explaining historical biological phenomena by evolution must be expanded to encompass at least a twofold hierarchy of levels of biological causality, two kinds of units of selection—genes and species.

The ultimate proof, however, must come from hard evidence. Until a compelling evidence is found, species selection will be nothing but an explanation in search of phenomena to explain—very much like the theatrical characters in search of an author who would write about them in the famous drama by Luigi Pirandello. Since such evidence is very hard to produce, however, and since the absence of firm evidence cannot, and should not, make us to reject species selection as a heuristic device, I predict that it will haunt us for a long time to come.

MEGAEVOLUTION

10 Is Megaevolution More Than Historical Contingency?

For the great neodarwinian George Gaylord Simpson (1944), macro-evolution was the origin of evolutionary novelty and megaevolution was a subset of this, namely the origin of new families and even higher taxa. When he later argued for dropping these two terms altogether, Simpson (1953) did so largely on the grounds that this distinction is entirely arbitrary because all higher taxa first appear as new species, due to the microevolutionary process of speciation, and because there is no demarcation line to be drawn between these and whatever other categories of this sort might still be invented. Macro-evolutionary phenomena, however, are now—and certainly in this book—defined much more broadly, as all supraspecific patterns in space and time, with the largest scale phenomena being recognized as megaevolutionary ones (Hecht and Hoffman 1986). This modern distinction between micro-, macro-, and megaevolution is not necessarily causal (whether or not it is causal and points to different biological processes is the bone of contention in many current arguments on evolution and, in fact, is the subject of this book), but it is not necessarily arbitrary either, even though the differences between these kinds of historical biological phenomena are related primarily to their scale.

Microevolution is what goes on within populations and species. Macroevolutionary concepts—punctuated equilibrium and species selection—are intended to provide explanations for such phenomena as distribution of rates along an evolving lineage, evolutionary trends within a monophyletic taxon, deployment of species in the course of evolution of individual systematic groups. These phenomena can be described and discussed in terms of either the transformational or the taxic approach to historical biology. They can be explained within an evolutionary conceptual framework with either the individual organism, or species as the benchmark of evolution. When we move up the spatial and temporal scale of analysis and up the hierarchy of biological organization, however, we shift our focus from macro- to megaevolutionary patterns and consider changes in structure of the biosphere, or at least of the biota inhabiting a certain ecological realm—for example, marine benthos, pelagic plankton, land plants. At this point, the approach must become taxic simply because the structure of these biological entities can only be described in the fossil record in terms of species numbers.

In principle at least, the structure of the biosphere could also be described in terms of energy and matter fluxes through its various components. This approach may in fact be crucial if we are ever to understand the functioning of the global ecosystem. Such a description, however, even if it is entirely appropriate for the living biosphere or biota, is obviously unachievable for the geological past. Hence, changes in structure of the biosphere cannot possibly be considered in terms of energy and matter fluxes on the geological time scale, which is the proper focus of megaevolution. They must be treated from the taxic perspective, expressed by changes in number of various kinds of species or even higher taxa. Species or clade is the benchmark of megaevolution.

The study of megaevolution differs therefore from macroevolutionary investigations not only in its focus—the kind and scale of the patterns to be recognized, described, and explained—but also in the approach—the mode of pattern description and, ultimately, explanation. Insofar as macroevolutionary phenomena are concerned, macroevolutionary explanations compete with microevolutionary ones as rival hypotheses to be tested against empirical data and compared to each other. There exist macroevolutionary patterns for which both kinds of the explanation are conceivable. An evolutionary trend in a phyletic lineage, for example, can be explained either by the prevalence of directional selection, even if variable in force and direction, or by species selection. A punctuated pattern of morphologic evolution in a lineage can be explained, for instance, by a considerable variation in phenotypic evolutionary rates due to the gradual appearance of developmental canalization and its subsequent breakdown, or by the law of punctuated equilibrium which associates all morphological change in evolution with speciation. In both cases, the former explanation is microevolutionary—in the sense of being a direct consequence of the genetical theory of evolutionary forces—whereas the latter is not. The former fits the neodarwinian paradigm, the latter does not because it envisages evolutionary forces operating upon species as units rather than within populations and species. As I have argued in the preceding chapters, there is no evidence to support the claim that macroevolutionary phenomena demand macro- rather than microevolutionary explanations; that they must refer to processes with the species rather than the individual organism being the benchmark of evolution.

The situation is very different with megaevolutionary phenomena. The genetical theory of evolutionary forces does not, and cannot, predict anything about patterns of change in structure of the biosphere; it does not, and cannot, tell anything about patterns of change in number or relative proportions of various kinds of species, and even less so for higher taxa. According to the neodarwinian paradigm of evolution, such patterns result from the incessant interplay between microevolutionary processes and historical contingency—the accidents of life on the Earth. In this view, megaevolution is historical contingency because history rather than any law of biological evolution determines the kind and severity of geochemical, climatic, geographic, and oceanographic changes that occur and affect the biota, which taxa will encounter opportunities for adaptive radiation upon invading by chance a new continent or ecological realm, which taxa will undergo divergent evolution due to dramatic fragmentation of their previously continuous habitat, and which ones will meet new predators and competitors due to habitat coalescence by continental drift. In other words, the neodarwinian paradigm envisages megaevolutionary patterns of biotic diversification, explosive radiations, and mass extinctions as reflecting the sum total of individual responses of myriads of species to a plethora of environmental challenges; these responses are the effects of a variety of microevolutionary processes, including species extinction when no adequate evolutionary response is possible. What the paradigm does not envisage, and cannot in fact accommodate, are megaevolutionary patterns that are to be explained by specific megaevolutionary laws; for the existence of such patterns, or rather the necessity of such explanations, would imply the operation of some evolutionary forces beyond those operating within populations and species and considered by the genetical theory.

Thus, there exist no evolutionary patterns that could be subject, even if only in principle, to both micro- and megaevolutionary explanations. For megaevolutionary phenomena, the competing hypotheses to be tested against empirical data and compared to each other are megaevolutionary explanations and historical contingency. The rival hypothesis is what makes the difference between macro- and megaevolution—as I understand here these terms—nonarbitrary.

Many of my fellow paleobiologists undertake to explain various megaevolutionary phenomena—mass extinctions and the pattern of taxonomic diversification in the Phanerozoic stand out as the most spectacular examples of such patterns—and there are many who regard these explanations as challenges to, or even refutations of, the neodarwinian paradigm. The question that must then be asked is whether these patterns reflect indeed anything other than historical contingency superimposed on the effects of microevolutionary processes going on in individual species.

11 Mass Extinctions: No Counterargument to Neodarwinism

11.1 A Bit of History

Death naturally belongs to the individual organism's condition, and it has always been perceived by scientists as the inevitable end of individual life. Species extinction, however, does not appear as ineluctable as individual death. Logically, there is nothing unavoidable about it, no matter whether or not one accepts evolution as the way of species origination. In fact, until the very end of the eighteenth century, natural scientists did not believe that species can really become extinct. But since Georges Cuvier (1799) demonstrated for the first time in the history of science the reality of species extinction, questions about the tempo and mode of extinction, and consequently about its causation, have been considered parallel to, and usually independent of, questions about species evolution.

The evolutionist Jean-Baptiste Lamarck (1809), for example, regarded species as essentially immortal because they were fully capable of evolution by adaptation to the changing environmental conditions, and he hence attributed all species extinctions to the actions of man, who alone could have violated the order of nature. The nonevolutionist Giovanni Brocchi (1814) conceived of each species as having its own predetermined life span—as all individual organisms did in his view—and, hence, of each species becoming extinct when its time came, independent of extrinsic factors. The interpretation of species extinction as caused by either exhausting the species' life potential, or by approaching the end of its life cycle was later taken up by such prominent, though strongly anti-neodarwinian, evolutionists as Othenio Abel (1904) and Otto Schindewolf (1950), respectively. Abel belonged to the leading advocates of the orthogenetic theory of evolution, Schindewolf developed his own theory of macroevolution based on analogy between evolutionary history of a lineage and life history of an organism.

The anti-evolutionist Cuvier (1825), in turn, attributed all species extinctions to local catastrophes of unspecified nature that should have wiped out all life from the affected areas; such an area would then be repopulated by immigrants from other geographic regions, and so on. For Cuvier, these catastrophes were, and must have been, always local because he had no evidence to extend their scope beyond the area he directly studied (being an avowed empiricist he refrained from baseless extrapolations). Moreover, given the existence of life on the Earth past several such catastrophes, a sequence of true holocausts on the global scale would also demand a sequence of repeated acts of Godly creation, which Cuvier, a devoted Protestant in a Roman Catholic country, could not possibly allow for. It was the creationist William Buckland (1823) who advocated worldwide catastrophes as the causes of the prononunced breaks in the faunal succession, but he did so explicitly on a religious basis. Charles Lyell (1832), however, even as a creationist, before he was converted by Darwin to evolution, had considered the extinction of species as a part and parcel of the economy and order of the natural world. In his view, species cannot possibly be immortal because, given the inevitable alterations of the environmental conditions on the Earth's surface, each species must sooner or later encounter a truly hostile situation and hence perish one after another, like the individuals that compose them.

Thus, a wide variety of opinions were already represented before Darwin: no evolution and no extinction of species, evolution but no extinction, extinction—either gradual or catastrophic—but no evolution. Darwin himself complemented this spectrum of positions, by accepting both evolution and extinction of species. In his opus magnum, *The Origin of Species,* Darwin (1859) reasoned that species extinction was caused either by physically or biotically controlled fluctuations in the numbers of individuals and in the species' geographic range, or by outcompetition of species by their better adapted descendants. He thus envisaged a multitude of processes that could actually lead to species rarity and, consequently, to the ultimate extinction. This pluralist view of species extinction is reflected in the neodarwinian paradigm.

When taken in conjunction with the logical independence of extinction and evolution, this pluralism explains why extinction usually occupies so little space in neodarwinian books on evolution. There is very little to say about species extinction as a scientific problem, and the little there is to say belongs to conservation biology rather than to evolutionary biology. Perhaps the only generalization of more direct interest to evolutionists concerns the correlates of species duration on the geological time scale, which is equivalent to species resistance to extinction agents. It appears to be well established that the larger the geographic range of a species, or the larger the number of biogeographic provinces a species inhabits, the longer it survives. This rule has been abundantly corroborated for a wide variety of organisms (e.g., Jablonski 1980, Hoffman and Szubzda-Studencka 1982, Martinell and Hoffman 1983); assuming, of course, that the morphospecies recognized by paleontologists are indeed biological species, or at least that the frequency distribution of morphospecies durations among various ecological categories of organisms does not significantly differ from the actual species pattern. One might even refer this relationship between biogeographic range and longevity of species to megaevolution and claim that it demands a specifically megaevolutionary explanation. The explanation for this megaevolutionary pattern, however, boils down to a simple probabilistic argument—that all factors of species extinction are more or less restricted geographically and hence, the greater and more heterogeneous is the total area occupied by the populations of a species, the greater is also the likelihood that at least one population will remain unaffected by any single factor. Naturally, the same empirical generalization, and also explanation, applies to genera and perhaps even families.

A subset of species extinction phenomena are mass extinctions. These are strictly megaevolutionary phenomena. They are periods—relatively short on the geological time scale, but perhaps, according to some hypotheses, very short indeed, measured in no more than months or years—of considerably intensified extinction of species, encompassing simultaneously many taxonomic groups and perhaps many ecological realms, and leading to a marked decrease in the diversity of the biosphere. They make an obvious impact on the course of the evolution of life, and hence they are of great interest to the evolutionary biologist and, especially, paleobiologist.

The marketplace of scientific ideas has long been rich in interpretations of mass extinctions. Buckland might be regarded as the godfather of the concept of global mass extinction, a holocaust affecting at once all of the biosphere. Cuvier, by contrast, spoke only of regional mass extinctions. Darwin, too, felt at ease with regional catastrophes. In his *Geological Observations on South America,* he attributed the mass extinction of large land mammals in South America to the effects of adverse climatic conditions during the Ice Age which must have rapidly reached that area (Darwin 1846). Later, he also noted some "wonderfully sudden" extinctions of large taxonomic groups—for example, trilobites at the end of the Paleozoic and ammonities at the end of the Mesozoic—but he suggested that the apparent rapidity of these phenomena might actually be an artifact of stratigraphic gaps in the fossil record; for, if a substantial time interval is not recorded by sedimentary strata, a period of gradual demise of an organic group may be lacking, thus producing the appearance of its very abrupt, virtually instantaneous extinction (Darwin 1859). Darwin thus proposed the now obvious alternative to Cuvier's literal reading of the record.

The assessment of validity of Darwin's remark on imperfection of the fossil record has considerably varied among later students of mass extinctions and largely shaped the course of the further debate. Some workers considered the extinctions toward the end of the Mesozoic—the Cretaceous/Tertiary extinctions which encompassed dinosaurs, ammonites, a wide variety of benthic marine bivalves, gastropods, echinoids, and bryozoans, and especially calcareous planktic organisms of the open ocean—as an abrupt catastrophe and sought adequate, catastrophic causal factors in varying intensity of the cosmic radiation (e.g., Marshall 1928, Hennig 1932). Other workers, in turn, viewed the same extinctions as extended over a substantial period of time, several million years, and looked for extinction agents that could operate on that time scale: either climatic side-effects of long-term tectonic processes involving mountain-forming as well as vast marine transgressions and regressions, or major changes in the composition of the biosphere leading to intense competition with groups better adapted to deal with new prey, predators, and parasites (e.g., Pavlova 1924, Golenkin 1927, Sobolev 1928). Similar differences in opinion existed about the extinctions toward the end of the Paleozoic—the

Permian/Triassic extinctions which encompassed primarily brachiopods, corals, cephalopods, and crinoids, but also affected virtually all
other marine animal groups.

With increasing knowledge of relatively complete geological sections at the Permian/Triassic and Cretaceous/Tertiary transitions,
however, it appeared that the seemingly catastrophic nature of these
events could not be fully attributed to mere gaps in the record. At this
point, then, even Otto Schindewolf had to conclude that the major
biological turnovers could not possibly be explained away by accidental clustering of the ends of the life cycle in a great many individual
taxa. These apparently catastrophic megaevolutionary phenomena
called for truly catastrophic events as explanation, and Schindewolf
(1954a, 1954b) decided to invoke a wave of cosmic radiation, perhaps
due to the nearby explosion of a supernova star, to account for these
mass extinctions. This concept was immediately and very severely
criticized (Simon 1958, Stepanov 1959) as reflecting a "desperate
move," as Schindewolf himself put it, rather than a scientific hypothesis based on and testable by empirical data. For apart from the apparently catastrophic nature of the Great Dying, there was absolutely no
evidence to support Schindewolf's scenario—nor even a faint idea
whether such deadly waves of cosmic radiation did ever reach the
Earth in the history of life. There was no chance for Schindewolf's
concept to become established as the best available on the marketplace of scientific explanations for mass extinctions, although it had
quite a number of followers among paleontologists even in the late
seventies. It was as unacceptable in science, but as sound if you
prefer, as the hypothesis that these catastrophes were caused by accidental impacts of huge bolides on the Earth—a recurrent theme both
among scientists and science fiction writers in the second and third
quarters of our century. The situation seemed so hopeless that all
sorts of wild ideas abounded: that a mysterious change in seawater
chemistry poisoned the majority of marine organisms, or that dinosaurs died out because of intoxication, or even constipation, caused
by flowering plants that appeared and flourished in the late Cretaceous. Because of the absence of any empirical evidence, all these
purported mechanisms of extinction belonged to what is sometimes
called "the lunatic fringe of science" rather than to science itself.

The scientific community was therefore greatly relieved when the
record of the Permian/Triassic extinctions, the greatest of all in the
history of the biosphere, was shown to be fully compatible with the

notion of their gradual nature and the time span of several million years. This assessment was immediately extrapolated over the other periods of geological time that have been traditionally recognized for mass extinctions: the Cretaceous/Tertiary transition; the latest Ordovician when the majority of trilobites, cephalopods, and brachiopods went extinct; the Frasnian/Famennian transition in the late Devonian when the Middle Paleozoic coral-stromatoporoid reef communities disappeared as well as a lot of associated organisms; and the late Triassic when a wide variety of shallow-water marine animals, mollusks in particular, suffered a heavy extinction (Newell 1967). To be sure, mass extinctions did still call for an explanation—or perhaps explanations, since there have been no compelling reasons to believe that they were all caused by the same agent—but the explanation could now be searched for among rather mundane forces. More than that, it should be sought among earthly forces simply because these are the only factors that have thus far been demonstrated to affect the biosphere. In this widespread view, mass extinctions should be conceived as representing simply more of the same phenomenon as the so-called background extinction, that is, individual species extinctions that happen all over the geological time due to myriads of independent causes. Hence, the causes of mass extinction should also be regarded as nothing but more of the same earthly, both biological and physical, forces that ordinarily lead to species extinction. This is not to say that extraterrestrial forces of whatever nature could not possibly influence the biosphere, but only that firm evidence must be provided before their operation can be accepted.

Such evidence was lacking, however, whereas empirical support could be marshalled for a variety of rather ordinary mechanisms of extinction. Thus, Thomas Schopf (1974) identified a striking correlation between the pace of the Permian/Triassic extinctions and the decline in the total area of continental shelves inhabitable by shallow-water marine organisms of the time; this decline was due to the coalescence of all land masses into a single supercontinent at that time. The smaller the available area, the smaller the number of species it can harbor, as the familiar ecological generalization goes. Changes of the sealevel as the ultimate cause of mass extinctions have been championed by many scientists, Anthony Hallam (1981) being perhaps their most prominent representative today. On the other end of the spectrum, many scientists have always regarded global climatic changes—most notably, cooling leading sometimes to glaciation—as the prime causal factor of mass extinctions. The argument to support this interpretation is currently best presented by Steven Stanley (1987).

Neither marine regressions or transgressions, nor global climatic changes, however, are consistently associated with mass extinctions, and none of these explanations is adequate to explain all mass extinctions. This has been long known but is perhaps most convincingly shown by David Jablonski (1986b) in his excellent comparative analysis of the five mass extinctions. Global changes in seawater temperature—induced by a climatic change—cause a shift in the geographic pattern of climatic zones, and marine faunas generally are able to migrate along with their habitats. As a result of overall cooling, tropical faunas may then end up in a sort of cul-de-sac because they obviously cannot track their disappearing warm-water habitats any further. Very often, tropical organisms are indeed most heavily affected by extinction. Yet extinctions are by no means confined to the tropical belt. Moreover, given their protracted duration in time, and given also the apparently great potential of various organisms to rapidly adapt to changed temperature in the environment, climate alone seems to be no more than "a half of the story," as Schopf (1984) once put it. Marine regressions, on the other hand, are not in themselves a mechanism of extinction. By decreasing the area of shallow-water continental shelves they presumably lead to increased interspecific competition and decreased average population size; they thus provide conditions for facilitating extinction of shelf faunas due to accidental vagaries of the environment. Jablonski observes, however, that marine regressions cause an increase in shallow-water area around oceanic islands, and since the island biota comprise a substantial proportion of the world's shallow-water marine benthos—close to 90% of the world's total family number today (Jablonski and Flessa 1986)—it is not at all clear if a major regression could induce enough extinctions on the shelf to produce impression of a pronounced mass extinction.

Despite these problems with finding an appropriate explanation for mass extinctions, however, even Digby McLaren's (1970) Presidential Address to the Paleontological Society—where McLaren argued for great rapidity and even instantaneity of the Frasnian/Famennian extinctions and for their causation by a large extraterrestrial impact on the Earth—could not change the widespread consensus about relatively protracted nature of mass extinctions and about their causation by earthly factors. There simply was no evidence for the action of such a cosmic mechanism, and the biostratigraphic correlation was too imprecise to determine instantaneity of the Frasnian/ Famennian extinctions all over the world.

This is not to say that McLaren's catastrophist scenario was regarded as wrong but only that there was no reason whatever to accept it as correct. There was no reason to view the Frasnian/Famennian extinctions as due to a catastrophic event, nor even as instantaneous. That an event is not instantaneous does not mean, however—even on the geological time scale—that it in fact spans millions of years, as it seemed to be the case with the Permian/Triassic extinctions. The late Devonian extinctions certainly happened much faster; and it has been known since the mid-sixties that the Cretaceous/Tertiary extinction among the pelagic plankton took place very rapidly and nearly isochronously wherever it could be studied in detail. With the advent of magnetostratigraphic techniques, basing time correlation of rocks on reversals of the polarity of the geomagnetic field, the span of Cretaceous/Tertiary extinctions in the pelagic realm could even be given some numerical value. Denis Kent (1977) estimated that these extinctions occurred within a time period that could not last longer than half a million years. This is very rapid by geological standards and it even allows for a real environmental catastrophe as explanation for the extinctions. An elegant scenario was in fact proposed that would fit into this time frame. Stefan Gartner and J. Keany (1978) suggested that a spillover of cold water from the isolated Arctic basin to the world ocean had happened at the Cretaceous/Tertiary transition and thus caused mass mortality among the pelagic plankton. According to Dewey McLean (1978), such mass mortality, in turn, would liberate tremendous amounts of carbon dioxide that should rapidly build up in the atmosphere and produce a greenhouse effect—very similar to the one now predicted to be the outcome of rising human production of carbon dioxide—considerably increasing the temperature and leading to extinctions on land as well as in the sea. Gartner and Keany's hypothesis fell on micropaleontological data, and McLean's scenario does not seem to allow for production of carbon dioxide in sufficient quantities to trigger very substantial environmental effects. But even their acceptance would not violate the consensus about the relatively gradual nature of mass extinctions and about their earthly causation. Obviously, it would also not prompt any calls for a revision of the neodarwinian paradigm which views all megaevolution—and mass extinctions in particular—as due to historical accidents of life on the Earth.

11.2 The Challenge

The situation has dramatically changed in the eighties. Luis and Walter Alvarez and their coworkers (1980) found an extraordinary concentration of iridium in the Cretaceous/Tertiary boundary clay layer at very distant locations. The clay layer is thin and as nearly isochronous as is possible to establish in geology—both by biostratigraphic and magnetostratigraphic means. It is not possible to rule out the hypothesis that it is in fact isochronous wherever it occurs, and it actually occurs over a huge part of the world. Iridium is a very rare element in the Earth's crust but it is fairly common in many meteorites. Hence, the Alvarez team quite naturally concluded that the iridium anomaly was isochronous and reflected the geological record of an extraterrestial impact. A simple calculation of the amount of iridium deposited, assuming the anomaly was global and came from a single impact, suggested that the presumed meteorite would have to be very large indeed, on the order of 10 kilometers in diameter. Astronomers estimate that the Earth should collide with a celestial body of this size approximately every one hundred million years or so. An impact of such dimensions should of course have tremendous—and instantaneous on the geological time scale—environmental consequences. The initial shock and heat wave would be followed by a dense cloud of dust being raised into the atmosphere and transported by winds all over the Earth; it would cut the biosphere off from the influx of solar energy, leading to freezing temperatures, a breakdown of photosynthesis, and a collapse of many ecosystems. In short, the nuclear winter scenario would have been realized in the geological past. Since the presumed impact should have coincided in time with the Cretaceous/Tertiary extinctions, and certainly with the extinction event among the calcareous plankton of the open ocean, it was logical to propose that it had really caused this mass extinction.

Soon afterward, the Alvarez team (1982) and Ganapathy (1982) found an iridium anomaly associated with the Eocene/Oligocene boundary which has long been regarded as a mass extinction, though never with such certainty as in the case of the five main revolutions in the history of life. This finding considerably reinforced the hypothetical link between extraterrestial impacts and mass extinctions, especially since the Eocene/Oligocene transitional strata also been known to bear microtektite fields, while these glassy microspherules are universally considered as definite evidence for meteorite impacts. Therefore, when David Raup and Jack Sepkoski (1984) discovered that their thorough statistical analysis of the distribution of marine animal extinction intensity per geological stage over the last 250 million years of the history of the biosphere suggested a periodicity of extinction peaks—including the Cretaceous/Tertiary, Late Triassic, and Permian/Triassic mass extinctions as well as the Eocene/Oligocene transition—it was logical to search for explanation of this emerging megaevolutionary pattern among astronomical mechanisms that could provide for a more or less regular influx of potential impactors into the neighborhood of the Earth's orbit. To do so seemed particularly logical since neither Raup and Sepkoski, nor anybody else could suggest a biological or other earthly mechanism responsible for a periodicity of extinction peaks on the order of 26 million years. A period of this length simply appears too long to be accounted for by ordinary evolutionary processes.

Raup and Sepkoski were very careful to repeatedly emphasize in their later writings the absence of a necessary logical link between the proposed periodicity of extinction peaks and their supposed extraterrestrial causation. The Cretaceous/Tertiary event, and also the alleged Eocene/Oligocene one, might merely coincide by chance with a peak of the periodic pattern. At first glance, nevertheless, the concept of geologically instantaneous mass extinctions caused by periodic extraterrestrial impacts certainly appears as the most parsimonious conclusion. For if extinction peaks occur periodically in the history of life on the Earth, they all are most likely to be due to a single causal mechanism; if this periodic series of events includes all mass extinctions that occur in the considered time interval, the causal mechanism in common is most likely to be what actually distinguishes mass extinctions from the background extinction; and if there is no plausible earthly mechanism that could account for a 26-million-year periodicity of historical biological phenomena, whereas two of these periodic events, the best studied ones, are indeed caused by asteroid or comet impacts and their direct aftermath, they all are most likely to be due to such cosmically controlled catastrophes. Given, therefore, that all these if-clauses are correct, the concept of mass extinctions by periodic extraterrestrial impacts is worth serious consideration by paleobiologists and also by astronomers, who are challenged to explain this megaevolutionary pattern by providing an adequate astronomical mechanism.

Astronomers, in fact, found an impressive array of mechanisms that could account for the approximate 26-million-year periodicity of impacts and, consequently, the periodicity of extinctions: A twin star of our Sun, christened poetically Nemesis, could periodically disturb the Oort comet cloud and direct an increased number of comets toward the inner planets of the solar system; the undetected tenth planet of our solar system, often dubbed Planet X, could travel beyond the orbit of Pluto and disturb the comet cloud in a similar manner as Nemesis; oscillations of the solar system about the plane of the galaxy could also cause periodic disturbance of the comet cloud by molecular clouds concentrated near the galactic plane, with periodicity being on the order of 30 million years or so. None of these explanations for impact periodicity has thus far been corroborated by any evidence, and all of them have encountered more or less serious troubles, but none of them could be refuted either (see Sepkoski and Raup (1986) for references and discussion) and perhaps some additional plausible astronomical mechanisms will be proposed. In any event, it is perfectly conceivable that there is some periodicity in occurrence of extraterrestrial impacts—the statistical analysis of Sepkoski and Raup (1986) suggests in fact a periodicity in geological age of large impact craters very close to the periodicity they found for extinction peaks—and if there is any, then its causes must be sought in the sky rather than on the Earth. That such causes can be found has already been demonstrated.

This conclusion adds plausibility to the concept of mass extinctions by periodic impacts, because the precondition of an adequate astronomical causal mechanism is likely to be met. If, furthermore, mass extinctions are different from the background extinction in their biological effects—as suggested by David Jablonski (1986a) on the basis of his analysis of mollusk victims and survivors of the Cretaceous/Tertiary transition—then a general theory of mass extinctions as a separate class of megaevolutionary phenomena might be called for.

As put by Stephen Jay Gould (1985), such a theory of mass extinctions should go beyond the neodarwinian paradigm, because if mass extinctions are caused by periodic asteroid or comet impacts and are "more *frequent,* more *rapid,* more *extensive in impact,* and more *qualitatively different in effect*" than traditionally expected, the mi-

croevolutionary processes envisaged by the paradigm are inadequate
to determine the shape of the biosphere—the course of megaevolution. Gould earlier wrote of this new perspective on mass extinctions
as being conceptually analogous to punctuated equilibrium, whereby
he evidently meant what I call in this book the radically strong version
of punctuated equilibrium, or the claim for discontinuity as the decisive characteristic in the history of life and life forms on the Earth
(Gould 1984). This opinion is supported by David Raup (1986) and
Kenneth Hsü (1987), for whom the new concept of mass extinctions
represents a vindication of Cuvier's catastrophism, which was for a
century and a half suppressed by the Lyellian, and also Darwinian,
paradigm of gradual patterns and processes in the history of both life
and Earth. Raup feels that it is primarily the gradualist pro-Lyellian
and anti-Cuvierian prejudice that initially hindered him, and still hinders others, from accepting that mass extinctions could be caused by
periodic extraterrestrial impacts. Hsü does not believe in periodicity,
but he asserts that even if only the Cretaceous/Tertiary extinctions
were caused by the Earth's collision with a substantial celestial body
rather than by competitive superiority of the survivors, this fact alone
ultimately refutes the theory of natural selection and of course the
entire neodarwinian paradigm.

11.3   How Can We Distinguish Mass Extinctions?

All of these are rather strong and even provocative propositions, and
it is no wonder that they arouse much controversy. But before their
validity can be discussed, the evidence for all the components contributing to the new perspective on mass extinctions must first be evaluated. The first question to be asked must of course concern the ways
to distinguish mass extinctions. If there is to be a general theory of
mass extinctions, it must refer to a clearly defined—or at least
delimited—category of historical biological phenomena. Traditionally, the Late Ordovician, Late Devonian (Frasnian/Famennian),
Permian/Triassic, Late Triassic, and Cretaceous/Tertiary, and perhaps also Mid-Cretaceous and Eocene/Oligocene periods of intensified extinction are regarded as mass extinctions, but it is an entirely
open question whether this tradition represents anything but a firmly
established convention. One way to address this problem is to look at
the magnitude and biological qualities of extinction in various time

intervals, in order to see if these characteristics allow for distinction of a class of extinction phenomena that would include the traditional mass extinctions. The point is to seek a nonarbitrary way of making such distinction.

There is no good empirical data to study the variation in rate or intensity of species extinction in the geological past. The vagaries of the processes of fossilization and of the subsequent geological history of individual sedimentary strata and even whole depositional basins make the fossil record a very unreliable source of information on species extinction on the global scale in particular time intervals in the history of the biosphere. The vast majority of species that existed became extinct without leaving any record of their past existence, and hence without giving us any chance of taking them into account in our data on species extinction. The empirical data become more reliable, however, when we move up the taxonomic hierarchy—an extinct genus including several species has, on the average, more chances of being found in the fossil record than any of its constituent species, a family more than any of its genera, etc.

Jack Sepkoski (1982) made a compilation of the first and last records of all marine animal families in the history of the biosphere. The time frame he employed for this compilation consisted of geological stages—the finest stratigraphic units that could consistently be recognized as nearly isochronous all over the world. In other words, Sepkoski located, by means of stratigraphic time correlation, all the paleontological records of marine animal families within an appropriate geological stage of the Phanerozoic, or the last 600 million years or so, and he then identified the earliest and the latest records for each family. On this basis, he could later calculate the intensity of marine animal family extinction per geological stage, where extinction intensity is defined as the fraction of families existing within a time interval that actually go extinct within this interval. And given the absolute duration of each stage, in millions of years, he could also calculate the intensity and rate of family extinction per million year in each stage. It is only natural that the longer the duration of a time interval, the greater number of families go extinct within this interval; to account for this bias, extinction metrics per stage must be divided by stage durations. Intensity and rate of extinction per million year can thus be arrived at. Sepkoski's data set has rapidly become the main source of quantitative information about the variation in extinction rate and intensity in the geological past.

Judging from these data, there is no quantitative criterion to distinguish between mass extinctions and the background extinction in the Phanerozoic. Stephen Stigler (1987) demonstrates that the set of family extinction rates per million year is better described by a single random distribution than by a model with two discrete classes of phenomena; his result suggests that this set represents a single statistical population. Quantitatively, therefore, mass extinctions cannot be unequivocally distinguished in life's history on the Earth—at least at the presently available level of stratigraphic resolution. The latter qualification is indeed crucial because the calculations of extinction metrics per million year provide in fact only the average figures for each stage and hence make sense solely under the assumption that family extinctions occur continually and are distributed more or less uniformly throughout each stage. If the majority of family extinctions within a certain stage were actually clustered at a single moment in time—as it would be the case if the stage included an episode of instantaneous mass extinction by, for example, extraterrestrial impact—then these data would be internally incoherent. An obvious peak of extinction rate or intensity per million year would then be artificially flattened and stretched in time because its constituent family extinctions would be uniformly apportioned among each of several million-year-intervals; the longer is the particular stage and the more pronounced is the difference between mass extinction and the background level, the stronger is also this effect. Thus, it is possible that a quantitative distinction exists between the intensities of background and mass extinction—and it may even be demonstrated when higher-resolution data will have been collected and analyzed, though the genus-level data considered by Raup and Sepkoski (1986) per geological substage do not seem to support this supposition—but there is no evidence for it at the present state of the knowledge.

David Jablonski (1986a, 1986b, 1986c) analyzes the patterns of survivorship and extinction among late Cretaceous bivalves and gastropods from the Gulf and Atlantic Coastal Plain in North America during the last 15 million years or so of the Cretaceous. He observes—in full agreement with the biological common sense—that highly speciose genera and geographically widespread species and genera persisted on the average longer during background extinction times than did species-poor taxa and taxa with a relatively small geographic range. He finds the same difference also between species with planktotrophic and nonplanktotrophic larvae, which is not surprising given the natural and well-established correlation between the ability of planktic larvae to feed, their ability of long-range dispersal by oceanic currents, and the resulting longer geographic range of the species than in the forms whose larvae survive only a short period of floating as plankton. During the Cretaceous/Tertiary mass extinction of these mollusks—approximately 60% of the genera did not survive this transition—the pattern of survivorship and extinction changes, however, in that both speciose and species-poor taxa and both geographically widespread and geographically restricted species are equally vulnerable to extinction, and also that large geographic range at the genus level becomes the only biological characteristic that correlates with better chances of survival under the mass extinction regime. Jablonski compares these results to a handful of data available for the other mass extinctions and proposes on this basis that a clearcut qualitative difference exists between the two evolutionary regimes—background and mass extinction.

If this conclusion about a qualitative distinctness of survival and extinction patterns during mass extinction were valid for all the traditionally recognized mass extinctions, it could provide a criterion for discerning this class of megaevolutionary phenomena. Detailed studies of the kind Jablonski has done on the late Cretaceous and the Cretaceous/Tertiary transition are thus far unavailable for other mass extinctions, however. Perhaps more important, it is far from being established beyond any reasonable doubt—even at the current state of the knowledge—that this conclusion indeed holds even for this best studied mass extinction event. The agents of mass extinction are, of course, more global in scope and affect a wider spectrum of ecological categories of organisms than do agents of background extinction. It must be so even if we deal with a single agent operating with differential intensity; for the more global scope of extinction phenomena and the wider ecological spectrum of extinction victims are exactly what has led to recognition of the mass extinctions and hence must, by definition, be typical of each mass extinction. On the other hand, genera must have on the average a greater geographic range than their constituent species. There is consequently nothing particularly surprising in the fact that even the widest distribution of individual species—within a single biogeographic province—may not suffice to save species from extinction by a pronounced environmental disturbance, whereas the wider is the distribution of a genus, often reaching beyond the borders of one bioprovince, the better should be its chances to survive the same environmental change. The pattern of widespread genera being less affected by mass extinction than geographically restricted genera occurs also in the Late Triassic (Hoffman 1986).

More surprising might be the fact that the species-poor and species-rich mollusk genera analyzed by Jablonski were equally vulnerable to the Cretaceous/Tertiary mass extinction. More speciose genera could in principle be expected to cover a wider spectrum of ecological adaptations to various environmental regimes and hence to have a better chance that at least one of its constituent species will remain unaffected by the extinction agent. This expectation presumes of course the usual qualification of "all else being equal." It may not hold, however, with respect to some specific environmental parameters. It may also not hold if the considered species-poor taxa are substantially less specialized ecologically than the species-rich taxa to which they are compared. These may well be the reasons why, for instance, some species-poor groups of ammonites survived much longer than their highly speciose relatives. Furthermore, species richness will of course be entirely irrelevant to survival if one compares taxonomic groups that widely differ in their ecological requirements—mammals and mollusks, for example, or trilobites and brachiopods, or benthic and planktic foraminifers—because each of the groups compared will then respond to a different set of environmental parameters, and the environmental disturbance that is deadly for one group may leave the other group unaffected. Thus, the pattern observed by Jablonski may not be surprising after all, for although it covers a taxonomically and ecologically coherent set of bivalves and gastropods, the degree of specialization of individual taxa is unknown. More important, however, this pattern recorded among the late Cretaceous mollusks from North America contrasts with the results I have obtained for the coeval mollusks from Europe and the southern Soviet Union, where a higher proportion of species-rich genera survive into the Tertiary than is the case with species-poor genera (Hoffman 1986). It is not clear what biological or environmental factors have caused this difference in the pattern of survival and extinction. A greater ecological heterogeneity of the European evolutionary theater might play a role here, but reliable quantitative data to demonstrate a difference in habitat heterogeneity are rather hard to obtain for the geological past. No matter what these controlling factors are, however, this discrepancy between Europe and North America indicates that the North American pattern may be a local phenomenon.

Thus, the presumed qualitative difference between the evolutionary regimes of background and mass extinction may actually boil down to widespread genera, but not species, being less vulnerable to mass extinction than geographically restricted taxa. It is the same simple probabilistic phenomenon that is so well known for species extinction; it is only raised to a higher taxonomic level, perhaps beyond a certain threshold value of geographic range, because the extinction agent operates during mass extinction on a much greater geographic scale and on a much wider spectrum of habitats. This effect of elevating the taxonomic level of survival pattern due to the greater scope of mass extinction factors is analogous to the effect of prolonged attrition wars on life expectancy in various age classes of man. During peace time, life expectancy is age-dependent; past the period of high infant mortality and beginning with the early peak at children's age, life expectancy generally decreases with age. During a war, however, and especially during a very long all-out war when mobilization encompasses several age classes, the average life expectancy is approximately equal and perhaps lower for 20-, 30-, and perhaps even 40-year-olds than it is for, say, 50- or 60-year-olds, but it still is significantly higher than for 80- or 90-year-olds. The pattern of age-dependent mortality persists, although its onset is shifted toward older age classes. During mass extinctions, similarly, the general pattern of geographic range-dependent extinction persists, but it is hardly discernible or even invisible at small geographic ranges. Whether a species has the range of 120 or 1200 kilometers may have some importance for its survival under local catastrophes and geographically restricted extinction agents, but no importance whatsoever during mass extinctions affecting half the globe. But whether a genus occurs in only one bioprovince or on several continents may determine its survival even under the mass extinction regime.

This is not to say that species with less than very wide or even global geographic distributions are nonselectively affected by mass extinctions, regardless of their various habitats and adaptations. David Raup (1979) estimated that close to 95% of all marine animal species had become extinct during the Permian/Triassic extinctions. Given this staggering figure, the taxonomic composition of the Early Triassic biosphere could be regarded as shaped primarily by chance survival of taxa that merely happened to be in the right place at the right time, rather than by any biological characteristics of the victims and survivors. This sort of capricious selection of species must operate locally, or even regionally, at all times but if it dominated on the global scale during mass extinctions, this might constitute a qualitative difference between the regimes of background and mass extinction. As pointed out by Raup, however, the statistical method he employed to obtain his estimate assumed nonselective extinction. In reality, the Permian/Triassic as well as the other four traditionally recognized mass extinctions were strongly selective, concentrated in a few higher taxa while only marginally affecting some others (McKinney 1985, Hoffman 1986). In such a situation, Raup's method considerably overestimates the intensity of species extinction. My own rather conservative estimate, or perhaps merely a learned guess, based on the observed pattern of extinction selectivity during the mass extinctions suggests that nonselective background extinction may sometimes have affected nearly as much of the biosphere as did selective mass extinctions (Hoffman 1986).

The distinction between background and mass extinctions thus appears to be largely arbitrary. The variation in extinction rate and intensity in the Phanerozoic seems to be continuous—or at least, there is at the moment no reason to reject this assessment. This situation is well reflected in a recent definition by Sepkoski (1986), which states that, "A mass extinction is any substantial increase in the amount of extinction (i.e., lineage termination) suffered by more than one geographically widespread higher taxon during a relatively short interval of geologic time, resulting in an at least temporary decline in their standing diversity." Sepkoski describes this definition as being "designed purposely to be somewhat vague"—intended to include the five time intervals traditionally regarded as mass extinctions but leaving it open which, if any, other episodes in the history of life should also be considered mass extinctions. Based on the analysis of both genus- and family-level data on the marine animal fossil record (Raup and Sepkoski 1986), Sepkoski discerns "29 local maxima in extinction intensity that may represent anything from protracted intervals of high evolutionary turnover to catastrophic events of mass extinction," and he considers them all as potential candidates for mass extinctions by extraterrestrial impacts. Given the absence of either quantitative, or qualitative biological criteria to distinguish mass extinctions, however, one must then rely on their causal uniformity as the criterion for their demarcation as a separate class of phenomena in the history of life on the Earth. Such a causal uniformity would be indicated, for example, by periodicity of extinction or by its causation by extraterrestrial impacts.

11.4 On the Periodicity of Extinction

Raup and Sepkoski (1984, 1986; Sepkoski and Raup 1986) analyze the compendium of data on stratigraphic distribution of marine animal families and genera in the Phanerozoic in order to test for periodicity in temporal distribution of peaks of extinction intensity. Extinction peak is here defined as a local maximum, or a time interval with extinction intensity higher than in the immediately preceding and succeeding time intervals, no matter how much or how little this extinction really affects the biosphere. The height of the peaks above the background is irrelevant. These peaks of extinction intensity include all the traditionally recognized mass extinctions but also quite a number of other time intervals with extinction intensity only very slightly higher than in their neighborhood. Since Raup and Sepkoski are interested in the temporal distribution of these extinction peaks in real time, they consider an absolute geological time scale, measured in millions of years instead of merely in such stratigraphic units as stages or substages. Because of the correlation problems, however, their data must have been collected within the framework of the relative geological time scale, assigned to stages or substages, and only subsequently attributed a numerical value in millions of years— all families or genera going extinct within a certain stage being thus arbitrarily assigned the same date of extinction, as if they all went extinct simultaneously at the end of the stage.

Having thus located the empirical pattern of extinction peaks in geological time, Raup and Sepkoski can ask the question, whether the temporal distribution of these peaks is periodic, that is, whether it deviates from randomness toward periodicity, whether it is more regular than should be expected if it were entirely random. First, however, they have to determine how the pattern of extinction peaks should look if it were indeed random, that is, if the peaks occurred in time totally independently of one another. This is not a trivial problem because there is only one historical pattern of extinction peaks, and the processes that have brought it about cannot be repeated or experimented with. Raup and Sepkoski solve this problem by generating a whole set of randomizations of the historical pattern. They create by computer simulation a large set of patterns that comprise the same data on family extinction intensity per stage but shuffled at random so that each pattern has them in a purely accidental sequence. They can thus establish a random distribution of extinction peaks in time. By fitting a strictly periodic function to the empirical pattern and to the computer-generated random distribution and by comparing the goodness of the fits, they are now able to detect if any—and if so, then which—periodic function deviates significantly less from the empirical pattern than it does from the random distribution. A smaller deviation from the empirical pattern indicates that this pattern is more closely periodic, with respect to the considered period length, than it should be expected to happen by pure chance. For the last 250 million years, Raup and Sepkoski find this to be the case for periods in the neighborhood of 26 million years. In other words, the empirical distribution in time of the peaks of extinction intensity since the late Permian until now appears more regular, less deviant from the 26-million-year periodic function, than it should be expected.

This result suggests then a periodicity of extinction and a causal uniformity of extinction peaks. It is this result that, in conjunction with the hypothesis of asteroid or comet impact as the cause of the Cretaceous/Tertiary and Eocene/Oligocene extinctions, led to the concept of mass extinctions as a separate class of historical biological phenomena caused by periodic extraterrestrial impacts. It is this result that led, consequently, to the astronomical hypotheses of comet showers thrown periodically on the Earth by the Nemesis star, or by Planet X, or by regular encounters with interstellar molecular dust clouds. On the basis of the concept of mass extinctions by periodic impacts, in turn, a megaevolutionary theory of mass extinctions might be possible.

This concept is seductively beautiful and based on sophisticated statistics, but there are also very good reasons for serious skepticism. Raup and Sepkoski's focus is on time distribution of peaks in a series of points that reflect a sequence of steps up and down. According to their definition, a peak of extinction is each turning point in the sequence where extinction intensity first goes up from one stage to another and then goes down again, no matter how far up and down it actually goes; what counts is only the direction but not the magnitude of change. A neutral model for description of such a pattern—to be tested as a statistical null hypothesis—may be the symmetric random walk model, where we always have 50% chance that any given step will lead upward and 50% chance it will lead downward. To be precise, this model is not entirely appropriate because it assumes that the variation in extinction rate or intensity can go up and down to infinity; only then would the probability of going up from a very high level of extinction be equal to the probability of going down. The actual variation of extinction, however, is limited by zero and the total number of taxa in the biosphere. Nevertheless, given the huge diversity of the biosphere, the random walk model may offer a reasonable first approximation. Assuming this model, then, there is 25% chance that the next change in extinction intensity after a step up will lead downward again; in other words, there is 25% chance that any given step will turn out to be a peak in the sequence. On the average, one step in four should be expected to be a peak under the model of symmetric random walk. The predicted average spacing of peaks is a peak every fourth step.

The time scale that Raup and Sepkoski employ, the one established by Harland and coworkers (1982), assumes that the majority of stages—in fact, 60% of the stages, which cover jointly 70% of the time interval, analyzed for extinction periodicity—vary in duration no more than from 5 to 7 million years, with the mean duration being 6.2 million years for the whole time span from the late Permian until now. This figure is indeed "assumed," since the time scale is constructed on the assumption that there are no good radiometric data for very long time intervals within this period which are therefore simply divided into approximately equal units. This procedure is of course arbitrary and probably incorrect, since the known variation in stage duration in the last 100 million years is from 1 to 15.5 million years, but it is often accepted for the sake of convenience and simplicity. On this largely arbitrary geological time scale, the symmetric random walk model predicts that there should be, on the average, one peak of extinction in each 24.8-million-year interval, which figure does not significantly differ from the 26-million-year periodicity observed by Raup and Sepkoski.

The average spacing of peaks alone does not demonstrate that the pattern of extinction peaks could arise by a process approximating the symmetric random walk. The variation in the peak spacing is crucial. Random walk is very unlikely to produce a really regular pattern, with one peak occurring every fourth step in the sequence. When the spacing of peaks in the empirical pattern of extinction is calculated in terms of the number of stages that separate them rather than in millions of years—and this pattern is actually constructed this way, for it is dated in absolute time only afterward, on the basis of the assumed geological time scale—then it does not strike one as being surprisingly regular for a series of approximately 40 steps. In fact, the statistical variance of the waiting times between peaks of extinction is not significantly smaller than predicted by the random walk model.

It appears therefore to be very likely that the empirical pattern of extinction peaks studied by Raup and Sepkoski is merely pseudoperiodic, a random pattern deviating by chance more toward a periodic than toward clumped distribution, instead of being truly periodic and only blurred by statistical noise. As a matter of fact, the alleged 26-million-year periodicity is considerably weakened or even disappears completely if different time scales are assumed—within the actual range of radiometric dating errors—and if different metrics of extinction are taken into consideration—rate or intensity of extinction per million year rather than per stage (Hoffman 1985).

My proposal that the megaevolutionary pattern of marine animal extinction may actually reflect a random variation in extinction rates has aroused a lot of controversy, and also misunderstanding, due in part to its implications for the concept of periodicity and hence causal uniformity of extinction peaks—for, if the pattern can arise from randomness, there is no reason to regard it as periodic and no need to search for its astronomical explanations—but in part also due to my imprecise wording. What is at issue here is not the question of whether the empirical pattern of extinction peaks is strictly periodic during the last 250 million years or so. Evidently, it is not. Whatever geological time scale is assumed, the waiting time between some peaks of extinction is exactly 26 million years, but it is much more than that between some other peaks and much less between still others. Nor at issue is the question whether a random process, like the symmetric random walk, can lead to a strictly periodic pattern. It can, but the likelihood of such a result is very small and becomes infinitesimally small with increasing length of the pattern. This is quite clear. If one flips a coin and gets heads every other flip, this result should not arouse suspicion concerning the coin fairness if it occurs in a series of four, six, or perhaps even eight flips; but the suspicion should almost change into certainty about cheating if this regularity persists after 20 flips. Briefly, I do not propose that a random variation in extinction intensity during the last 250 million years has led to periodic spacing of extinction peaks. Instead, the real bone of contention is whether the actual, historical pattern of extinction intensity could only be obtained with a contribution of periodic signal, as proposed by Raup and Sepkoski, or whether the observed amount of regularity is so small that it could also result from a stochastic process, as proposed by myself.

There are some strong quantitative arguments to support my view. Jennifer Kitchell and Daniel Peña (1984) demonstrate that if the magnitude of extinction peaks, and not only their spacing, is also taken into account, a stochastic model can indeed lead to pseudoperiodicity that fits the empirical pattern even better than does the deterministic, periodic model. Kitchell and Estabrook (1986) further demonstrate that approximately 8% of their simulated 10,000 symmetric random walks of the same length as the sequence studied by Raup and Sepkoski produce patterns with their peaks spaced even more regularly than in the empirical pattern of extinction intensity. Sheldon Ross (1987) shows that up to 35% of such symmetric random walks are statistically indistinguishable from the empirical pattern. Elliott Noma and Arnold Glass (1987) demonstrate that the historical pattern of extinction has no more regularity than expected to occur in more than 5% of the series of numbers produced by a roulette. Edward Connor (1986) finds that Raup and Sepkoski's data may suggest not only one but a few different periodicities, with more than one period length, while there is no way of determining which, if any, of these periodicities is true rather than apparent. Stephen Stigler and Melissa Wagner (1987) show moreover that, given the nature of empirical data on extinction intensity through geological time, Raup and Sepkoski's statistical test is intrinsically biased toward the result they have indeed obtained—26-million-year periodicity of extinction peaks.

I think that these analyses provide sufficient support for my proposal that the historical pattern of extinction intensity over the last 250 million years deviates so much from strict regularity, or so little from the expectations under the assumption of randomness, that it may reflect stochasticity rather than periodicity. That a stochastic model can account for the empirical pattern of extinctions may then imply two different things. Either the pattern indeed results from aggregation of myriads of independent species extinction events in a very wide variety of different marine habitats and is therefore truly random, or it primarily reflects a statistical noise due to a lot of independent taxonomic, stratigraphic, and taphonomic biases; even if extinctions were indeed periodic, the regularity could be over-whelmed by noise.

This is not to say that periodicity of extinctions is ruled out but only that it is not at this point the best among currently available interpretations of the historical pattern. Perhaps some further analy-sis will show that the observed deviation of the empirical pattern from what should be expected under full randomness is indeed sufficiently large to exclude pseudoperiodicity as a valid hypothesis. I doubt it, though. It is also worth noting that apparently periodic patterns also occur in such domains of reality where true periodicity is very hardly thinkable. For example, the timing of major social restructuring events in Poland since World War II is almost perfectly periodic (they took place in 1944, 1956, 1970, and 1980–1981) certainly not less so than even the most periodic part of the extinction pattern. Yet this historical periodicity is not seriously believed by anyone to indicate the action of a single causal process that periodically triggers social upheavals in Poland. The next political change in my country may indeed happen in five years from today, but perhaps we shall have to wait much longer or perhaps it will happen next year.

Very much in the same vein, a fascinating observation is made by Colin Patterson and Andrew Smith (1987). They reanalyze the fish and echinoderm data in Sepkoski's (1982) compendium on marine ani-mals, which constitute jointly more than one-fifth of the family-level data set studied for extinction periodicity, and find that the majority of this subset of data are more or less incorrect. This is not particularly surprising because a single person cannot possibly have the expertise to evaluate the enormous literature on taxonomy and stratigraphic distri-bution of all marine animals throughout the Phanerozoic; Sepkoski did an admirable job and his compendium has served its purpose very well. What is really unexpected, however, is that it is the taxonomic and stratigraphic noise rather than good paleontological data on fossil fishes and echinoderms that gives the impression of closeness to period-icity. There is, I think, no way to make sense out of this observation other than that the empirical pattern of extinction peaks approaches by chance pseudoperiodicity.

11.5   Extraterrestrial Impacts and Mass Extinctions

The validity of the concept of extraterrestrial impacts as the common cause of mass extinctions crucially depends at this moment on the occurrence of true periodicity in the empirical pattern of extinction intensity. For, in the absence of direct evidence for asteroid or comet impacts at all the mass extinctions or extinction peaks, only a periodicity of extinction peaks, in contrast to pseudoperiodicity arising from randomness, could be taken to suggest a common causation of all these extinction peaks. Since two of these allegedly periodic peaks, the Cretaceous/Tertiary and Eocene/Oligocene ones, also coincide with geochemical and mineralogical evidence regarded as indicative of extraterrestrial impacts, it is entirely reasonable to interpret this purported common causation as impacts. Otherwise, one would have to resort to mere accidental coincidence, which certainly is not an elegant solution to a scientific problem.

As noted repeatedly by Sepkoski and Raup (1986, Raup 1986, Sepkoski 1986), a supporting argument may also come from the reported evidence linking other extinction peaks with presumed impacts. Iridium anomalies have been described in association with the Frasnian/Famennian, Permian/Triassic, Callovian/Oxfordian in the Jurassic, and also Precambrian/Cambrian transitions. The first two of these are thus related to mass extinction time intervals, the third one is recognized as a peak of extinction intensity in Raup and Sepkoski's analysis, and the last one is associated with a shift in carbon isotope composition of the rocks, which has been interpreted as indicative of a dramatic drop in biological productivity of the ocean and hence of a mass extinction event. These data might support the concept of mass extinctions by asteroid or comet impacts.

The evidence linking impacts to mass extinctions or peaks of extinction is far from compelling, however. The following summary is based on my own, admittedly critical, review article (Hoffman 1988). A very thorough reinvestigation of the Permian/Triassic boundary in Chinese sections, where the enrichment in iridium was first recorded, failed to reveal a significant iridium anomaly. The original report might perhaps be caused by sample contamination, as the analytical procedure of measuring iridium concentrations is extremely sensitive; the measurements are in parts per billion. The Late Devonian iridium anomaly appears to be only local and to represent enrichment by some kind of bacteria under very slow rate of sedimentation; it has been found only in cyanobacterial structures in Australia and—thus far at least—nowhere else in the world. Moreover, insofar as conodont correlations are a reliable, in fact the most reliable, time indicator in the Late Devonian, this anomaly seems to occur above the stratigraphic levels where the bulk of extinctions took place. The Callovian/Oxfordian boundary anomaly occurs in Poland and Spain, but it is again associated with cyanobacterial structures and its isochroneity is questionable, since it occurs in a geological formation of demonstrably variable geological age even within Poland itself. The so-called Precambrian/Cambrian boundary iridium anomaly occurs actually well within the Cambrian and its associated shift in carbon isotype composition of the rock is just one among several events of this kind within the late Precambrian and early Cambrian, while there is absolutely no reason to suppose that this particular shift, any more than the others, is indicative of mass extinction. Moreover, processes other than mass extinction can also lead to such a geochemical phenomenon.

In spite of intense multinational search for iridium anomalies or other impact fingerprints associated with other mass extinctions or peaks of extinction—in the Cambrian, late Ordovician, and Mesozoic—none has been found thus far. There is an anomalously high concentration of iridium in the Pliocene, in association with what is undoubtedly meteoritic debris, but without any apparent relationship to a major wave of extinctions. Thus, the Eocene/Oligocene and Cretaceous/Tertiary transitions still provide the only potential evidence for extraterrestrial causation of extinctions. This evidence, however, is not decisive either.

The Eocene/Oligocene boundary concentration in iridium is indeed associated with microtektite fields, which constitute a rather undisputable evidence of meteorites, and it also roughly correlates with what has been traditionally regarded as a major extinction event. When studied in detail, however, the fossil record clearly shows either a gradual biotic change, or a series of minor events—depending on the researcher's perspective—ranging in time over several million years from the beginning of the Late Eocene until the middle Oligocene, both in the marine realm and among land mammals. There are in fact several microtektite layers in this stratigraphic interval, unrelated to the iridium anomaly and marine extinctions, but the precise dating and stratigraphic correlation of all these events are very far from satisfactory. Furthermore, there is ample geological evidence for very strongly pronounced and widespread volcanism at that time, whereas volcanic eruptions bringing up the material from the mantle have also been observed to bear a lot of iridium and could be responsible for iridium anomalies in the sedimentary record.

In the Cretaceous/Tertiary boundary clay layer, the evidence for a large extraterrestrial impact is strong. It includes: iridium anomaly and other geochemical indicators found in a very large number of geological sections of land as well as marine deposits; quartz grains metamorphosed under extremely high pressures and temperatures, as it happens in the sites of nuclear explosions; enrichment in fluffy carbon particles suggestive of soot and hence of widespread fires, as predicted to arise in the aftermath of a huge impact; mineral microspherules that can be regarded as altered impact droplets, analogous to microtektites. All these mineralogical and geochemical characteristics of the sediment are moreover associated with the Cretaceous/Tertiary boundary defined by either marine micropaleontological or terrestrial paleobotanical data. On the other hand, the relative abruptness of pelagic plankton extinctions in the ocean, the pattern of their selectivity—with Arctic diatoms, able to survive very long periods of darkness, surviving also the Cretaceous/Tertiary boundary crisis—and the very sharp disturbance of land plant ecosystems exactly at this boundary strongly suggest a causal link between the extinction event and the impact.

Even the scenario that Cretaceous/Tertiary mass extinction is due to direct consequences of an asteroid or comet impact is far from being firmly established, however. Both the components of this scenario—the impact and the geological instantaneous extinction—can be debated. As pointed out by Charles Officer and coworkers (1987; Hallam 1987), the sedimentary record of the Cretaceous/Tertiary transition is also compatible with the hypothesis that it reflects the environmental effects of a tremendous and sustained volcanism rather than of an impact. And the fossil record of the associated extinctions can well be explained as caused by these effects, if only because they must include almost all effects that would follow an impact, although they would not reach such extremes as foreseen by the hypothesis of impact-triggered global wildfires to be followed immediately by nuclear winter.

All the geochemical peculiarities of the Cretaceous/Tertiary boundary clay in marine settings, including the iridium spike, can be satisfactorily explained by precipitation of metals dissolved in pore water migrating upward through the sedimentary strata. Such a precipitation would have to occur at the boundary between oxygen-rich chalks of the Cretaceous and the overlaying boundary clay rich in sulfides. It may not be accidental that the iridium anomaly does not occur in transitional Cretaceous/Tertiary strata where they are comprised exclusively of chalk, without the boundary clay layer. If the underlying calcareous deposits were even slightly enriched in iridium and associated metals due to prolonged volcanic activity—and there is much geological evidence for extraordinarily intense volcanism during perhaps half a million years in the latest Cretaceous—the boundary clay would quite naturally contain an unusual concentration of iridium. In conjunction with reduced sedimentation rate of the boundary clay, volcanism of this kind can also provide an explanation for all the mineralogical characteristics of the clay layer: shocked mineral grains, fluffy carbon, microspherules. The paleoceanographic anomalies during and as much as 1 to 2 million years after the Cretaceous/Tertiary transition disagree with all the models of the climatic and oceanographic aftermath of an impact event in the ocean, whereas no large impact crater of appropriate geological age occurs on land. Thus, even the hypothesis of extraterrestrial impact alone encounters troubles and rivals.

The case is much weaker, however, for impact as the cause of extinctions. As argued by Lowell Dingus (1984) on the basis of analysis of the completeness of geological sections at the Cretaceous/Tertiary transition, it is doubtful that we can determine in these sections whether the pelagic plankton extinction took place as rapidly as within 100 or rather as gradually as over 100,000 years. In any event, the extinctions of planktic foraminifers and calcareous coccolithophorid algae were not exactly synchronous even on this scale, with the forams having disappeared from the pelagic ecosystem significantly earlier than the coccoliths. Moreover, the iridium spike occurs in some places above the foraminifer extinction; elsewhere, it is demonstrably nonisochronous relative to other stratigraphic markers or replaced by a couple of iridium anomalies.

The fossil record of land vertebrates, both mammals and reptiles, is among the most contentious issues in the whole controversy over the Cretaceous/Tertiary mass extinction—perhaps because dinosaurs attract so much public attention. This record, however, is so incomparably worse than the pelagic record that it cannot presently be employed as a strong argument one way or the other. There is consequently no evidence to establish abruptness of extinctions on land, as demanded by the impact scenario, but no evidence to exclude it ultimately either.

Marine macroinvertebrates, such as ammonites, bivalves, gastropods, belemnites, echinoids, and bryozoans, were undergoing a major evolutionary turnover—including both extinctions and originations—beginning at least a couple million years before the end of the Cretaceous. Many bivalves, for example, which were until recently thought to have first appeared only in the Tertiary, are by now known to occur also in the latest strata of the Cretaceous. On the other hand, many groups disappear from the record before the Cretaceous/Tertiary boundary—inoceramid and rudist bivalves and ammonites are the classic examples—thus contradicting the notion of immediate and simultaneous extinction of the vast majority of the biota. It is often pointed out in this context that, owing to the vagaries of the fossil record, even an abrupt extinction event could actually be recorded as a gradual one because the fossils of various groups would be removed at random from the record before the true extinction event. In such a case, however, different groups should disappear abruptly here but gradually there and in a different sequence in different local sections—because the random effects of biases in fossil preservation would operate differentially in each local geological setting—whereas the pattern of macroinvertebrate disappearance below the Cretaceous/Tertiary boundary is largely consistent wherever it can be investigated. It thus seems at least very likely that many marine macroinvertebrate groups were declining more or less gradually over a long time interval, on the scale of hundreds of thousands or even millions of years rather than months or at most years as predicted by the hypothesis of mass extinction by one enormous impact.

This more protracted process of extinction might also explain the correlation between the geographic range of genera and their rate of survival across the Cretaceous/Tertiary boundary, which Jablonski (1986a) observes among North American mollusks. If the extinctions were due to the global effects of a huge impact, wide range should offer no, or at best very little, advantage for survival. Under the operation of more geographically restricted extinction agents, even if acting with extraordinary intensity and covering, over a certain time interval, virtually the whole planet, wide geographic range of a taxon should provide it with somewhat better chances for survival of at least some constituent populations or species.

It is of course perfectly conceivable that the dwindling macroinvertebrate groups ultimately fell victim to the dire consequences of an extraterrestrial impact at the Cretaceous/Tertiary boundary—as proposed by Alvarez and coworkers (1984)—but this would certainly be a coincidence of several different processes. The hypothesis of Cretaceous/Tertiary mass extinction by impact must also account for two other remarkable coincidences: the increased rate of intrusion of mantle-type rocks dated close to the Cretaceous/Tertiary transition; and the overwhelming correlation between evolutionary pulsations in the pelagic biota, including the Cretaceous/Tertiary extinctions, and the spreading rates of the seafloor which reflect geological activity of the rift structures. These coincidences, however, are quite easily understood given the idea of tremendous and prolonged volcanism being the prime causal factor of the Cretaceous/Tertiary extinctions. One could of course hypothesize that the Earth's collisions with large extraterrestrial bodies are what triggers such extraordinary volcanism—for it is definitely not unthinkable that a strong impact can pierce the Earth's crust and reach the mantle—but there is no independent evidence whatsoever to corroborate such hypothesis. In any event, the impact would then have to occur at the onset of volcanic activity, which apparently precedes the Cretaceous/Tertiary extinctions of the plankton by several hundred thousand years.

Thus, the concept of mass extinctions by asteroid or comet impacts, which adherents believe avoids the need to invoke mere coincidence between impact and one of the periodic extinction peaks, relies itself upon the assumption of a number of coincidences. When coupled with the absence of compelling evidence for periodicity of extinctions, this conclusion fully justifies, I believe, my skepticism about the common causation of mass extinctions as the criterion to distinguish them as a separate class of megaevolutionary phenomena.

11.6   Mass Extinctions as Clusters of Events

There is no evidence that mass extinctions, or the other peaks of extinction intensity, are caused by the same factor or factors. There is in fact no evidence that mass extinctions are single events, global in scope and geologically instantaneous in time—that is, below the resolution potential of global stratigraphic correlation—rather than clusters of causally independent, temporally or spatially isolated episodes. In many instances, the weight of the empirical data may actually point to the latter possibility.

As summarized above, the Cretaceous/Tertiary extinctions among the pelagic plankton may qualify as catastrophic, though their apparently consistent temporal sequence might undermine even this assessment; but there certainly are very good reasons for skepticism about the catastrophic nature of extinctions among marine macroinvertebrates and on land. Donald Prothero (1985) and Erle Kauffman (1986) summarize data to show the composite, aggregate nature of the so-called Eocene/Oligocene extinction event. The aggregate nature of the Mid-Cretaceous extinctions is abundantly documented by Kauffman (1984, 1986). Some of them are closely related to, and probably caused by, an oceanic anoxic event, when the oxygen-poor water gets onto the shelf and changes the shallow-water habitats to anaerobic. Such an event may be caused by rapid marine transgression onto wide coastal plains when the seawater rapidly receives vast amounts of organic matter, or it may be due to ocean overturn when bottom water from the deep ocean gets to the surface. Such paleoceanographic events may be triggered by rapid climatic changes, sealevel changes, or tectonic movements in the lithosphere. Other extinctions in the Mid-Cretaceous seem to be clumped into distinct clusters spread over several million years, without any evidence for their global isochroneity.

Piet Hut, Walter Alvarez, and their coworkers (1987) argue, on the basis of the evidently aggregate nature of these three periods of intensified extinction, that mass extinctions are caused by comet showers or multiple extraterrestrial impacts spread over a considerable time interval. Their argument, however, is built entirely on the Eocene/Oligocene case, where several microtektite fields and two iridium layers are known, thus suggesting indeed multiple impacts, and on the evidence that neither the Cretaceous/Tertiary, nor the Mid-Cretaceous extinctions can be regarded as single catastrophes. They present no evidence whatsoever to connect the Mid-Cretaceous and late Cretaceous extinctions with any impacts; and the extinctions associated with, and allegedly caused by, Eocene/Oligocene impacts include episodes when no more than five out of approximately 300 radiolarian species went extinct. In my view, the evidence put together by Hut, Alvarez, and others suggests only that a single, even as huge as the presumed Cretaceous/Tertiary, impact is not enough to cause a mass extinction—simply because mass extinctions involve more than one event.

My assertion is supported by evidence from other periods of intensified extinction. As demonstrated by Anthony Hallam (1986), the

Early Jurassic peak of extinction spans at least two ammonite zones, thus extending presumably over several hundred thousand years, and clearly is a regional phenomenon, well reflected in Europe but not at all in South America. The causes of the early extinctions contributing to this peak are thus far largely enigmatic, but the latest episode of extinction is evidently related to an oceanic anoxic event. The Late Triassic extinctions of marine fauna appear to be relatively rapid and related to a major regression; on land, however, Michael Benton (1987) indicates their subdivision into at least two major episodes. For the late Ordovician extinctions, P. J. Brenchley (1984) shows a sequence of extinction episodes spread over at least 2 to 3 million years during the latest Ordovician and earliest Silurian stages. This cluster of extinctions clearly coincides in time with a major glaciation, and hence a causal relationship to climatic oscillations or rapid sea-level fluctuations induced by alternating glaciation and deglaciation phases appears self-evident, but the real nature of extinction mechanisms is far from obvious. In any event, the Pleistocene glaciation and its consequences did not lead to any major wave of extinctions. This clearly indicates that glaciation is not enough to cause a mass extinction. The sequence of geological events during the Late Devonian extinctions suggests that the Middle Paleozoic reef biotas were exterminated by the onset of oceanic anoxic regime, caused by a very rapid and extensive transgression of the sea, but the pelagic fauna survived this crisis and went extinct only a couple of conodont zones later, probably due to some enigmatic paleoceanographic factors (Narkiewicz and Hoffman MS).

The Permian/Triassic extinctions are the most mysterious of all, perhaps because their record is extremely poor, due to a great marine regression. They coincide in time with very pronounced climatic changes and with a dramatic loss of biogeographic provinciality—the number of marine bioprovinces seems to be smaller than at any other time in the Phanerozoic—and these factors could account for a major but protracted extinction. The Permian/Triassic transitional strata comprise in fact a mixed fauna, some typically Permian brachiopods and ammonoids coexisting with typically Triassic bivalves, and there is no evidence for an abrupt biotic change (Sheng et al. 1984, Yin 1985). On the other hand, recent studies on stable isotopes of carbon, oxygen, sulphur, and strontium (for example, Holser et al. 1986, Gruszczyński and Małkowski 1987) indicate a widespread and profound paleoceanographic change at the end of the Permian—more dramatic than anything recognized thus far in the remainder of the Phanerozoic—thus suggesting a possibility of relatively rapid and essentially global event.

Thus, apart perhaps from the Permian/Triassic extinctions, which may well be, at least to some degree, due to a single event that was global in scope, all the other major peaks of extinction appear to represent clusters of separate events, more or less accidentally aggregated together. Some of these clusters seem indeed to involve global, possibly but not necessarily catastrophic extinction events. Some extinctions seem to be caused by climatic or paleoceanographic changes—such as oceanic anoxic events and pulses of global cooling—others may be due to volcanism, extraterrestrial impacts, and whatever other agents can affect large segments of the biota at once. All these factors certainly operated repeatedly during the Phanerozoic. Extensive meteorite craters bear witness to several large extraterrestrial impacts; vast amounts of volcanic rocks, which originated very rapidly on the geological time scale, point to episodes of massive volcanism; geochemical, geological, and paleontological evidence clearly indicates a substantial number of various oceanographic and climatic events. Yet not all large impacts, huge volcanic eruptions, profound oceanographic and climatic changes—perhaps not even a majority of them—are linked to disastrous mass extinctions, or even to smaller peaks of extinction intensity. In general, it apparently takes more to effect a mass extinction than just a single event of any kind. Mass extinctions may rather reflect those rare instances in the history of life when the actions of several independent extinction agents happen to roughly coincide in time, thus adding together their individual and generally moderate effects to make jointly for a major "catastrophe."

There still is a lot to learn about all these extinction events; and, no doubt, there is more than enough room for controversy and arguments about mass extinctions. But it seems very unlikely that mass extinctions indeed are, as Gould (1985) put it, more frequent, more rapid, more extensive in impact, and more qualitatively different in effect than traditionally envisaged by evolutionary paleontologists. What is and what is not called "mass extinction," depends entirely on an arbitrary decision, and it is this decision that determines the frequency with which mass extinctions are believed to have occurred, and the role they are believed to have played, in the history of life. There is no empirical support for the notion that mass extinctions are distinctive in any qualitative way—due to either their common causation, or their similar effects on the biosphere. Every mass extinction, and every peak of extinction intensity, appears to be unique, both in its causes and in its effects. Thus, a general theory of mass extinctions is likely to be impossible, and the new megaevolutionary perspective on mass extinctions does not appear to be well anchored in geological and paleontological evidence.

Yet even if this new perspective were indeed valid, it could not possibly demand any profound change in the paradigm of evolutionary biology and paleontology. Given this perspective, the megaevolutionary phenomenon of mass extinctions would be explained as caused by some extrinsic, nonbiological agent, and whatever regularity there might be, it would be derived from the properties of this agent. Hence, a theory of mass extinctions—even if possible, which does presently not seem to be the case—would only mean that the pattern of megaevolution, or the history of life on the Earth, is not fully determined by biotic factors alone, but that it also depends, perhaps even crucially, upon other historical factors.

Species extinction in general, and mass extinction of species in particular, is a factor that cannot be neglected in any serious attempt to understand the present shape and the past history of the biosphere. Whether aggregated and gradual or singular and catastrophic, mass extinctions certainly played a considerable role in megaevolution. This uncontroversial assertion implies a challenge to study them carefully. It does not imply, however, the operation of any specifically megaevolutionary forces; it is not synonymous with acceptance of any laws of megaevolution. It only denies to life a complete autonomy from the environment. The neodarwinian paradigm, however—and for that matter Darwin himself—does not demand such an autonomy of life from its environment. Just the opposite is true. Within the neodarwinian conceptual framework, even microevolution is always an effect of the continuing interplay between these two complexes of causal factors, and it is even more so with megaevolution.

If, on the other hand, mass extinctions are in fact clusters of causally independent events—as I believe the majority, if not all, of them are—their occurrence in, and their effect on, the history of the biosphere become primarily dependent on historical contingency—accidentally joint action of a variety of factors. This is exactly what megaevolution is all about according to the neodarwinian paradigm.

12   Biotic Diversification:
A Dilemma of Unsatisfactory Data

12.1   The Debate about Paleontological Data
on Diversification

The compendium of the temporal ranges of marine animal families in
the history of the biosphere that Jack Sepkoski (1982) compiled, and
which is widely employed for analysis of extinction patterns through
geological time, was originally intended to—and generally does—
serve the function of the main quantitative data source for studies on
the megaevolutionary pattern of changes in global organic diversity
during the 600 million years or so of the Phanerozoic. The latter pattern should be ideally considered at different taxonomic levels, in
terms of global species richness, as well as genus, family, and order
richness, per absolute time unit. For what matters for the historical
biologist is not only whether the organic diversity has changed through
time, and if it did, then along which trajectory. It also matters how
these possible changes in diversity have been accomplished, how the
process of global biotic diversification occurs. The question can be, for
example, if an increase in diversity occurs primarily due to species
diversification within such large taxonomic units as families or even
orders, corresponding to very general body plans and ways of life, and
hence to relatively large segments of adaptive ecospace; or if it is rather
due to origination of new families or orders, that is, to appearance of
major evolutionary innovations allowing for occupation of new adaptive zones. In the former case, the family level richness might remain

roughly constant through time but the species richness would considerably increase, whereas in the latter case, the trends in family and species richness would parallel each other. It is via such questions that the problem in causation of the pattern of biotic diversity through geological time should be addressed.

This approach to the problem of biotic diversification, however, is hardly workable in the fossil record because a large number of sampling biases conspire to make the paleontological data an unreliable source of information on changes in global species richness through time. The fossilization potential of marine animals is, on the average, much higher than is the case with either land animals or plants. The difference is so striking that paleontological data sets on these three realms of biological reality appear to be incomparable; they have to be analyzed separately. The bulk of current debate on the megaevolutionary pattern of changes in global organic diversity is in fact confined to the marine animal record only, because it is so much better than anything else suited for this purpose. It is understood that any extrapolation to the level of the entire biosphere can be made only with much caution. Even the marine record, however, is very strongly distorted by a variety of sampling biases, the awareness of which seems to belong by now, after Simpson (1960), Cutbill and Funnell (1967), and Raup (1972), to the paleontological folk wisdom.

The volume of sedimentary rocks preserved and available for paleontological search for fossils strongly varies with geological time and generally decreases as we move backward in time, toward older rocks. This is not at all surprising because older rocks have had much greater chances of being subducted back into the Earth's mantle, metamorphosed, eroded, or simply covered by very thick layers of younger rocks which make them inaccessible for paleontological investigations. The observed species richness in the fossil record should of course parallel this trend, because the greater the amount of rocks studied, the greater the sample of fossils and the better the chance to encounter a new species. Strata of some geological ages include also *Fossil-Lagerstätten,* those exceptional accumulations of normally unfossilizeable ancient organisms, such as the famous Cambrian Burgess Shale in Canada, the Carboniferous Mazon Creek concretions in the United States, or the Devonian Hunsrück Slate and the Jurassic Solnhofen Limestone in Germany. These of course inflate the observed species richness in their respective time intervals relative to the segments of the fossil record where no comparable occurrences have been found.

The so-called "pull of the Recent," or the effect our excellent knowledge of the living biotas compared to those from the geological past has on paleontological interpretations, tends to increase organic diversity in younger rocks. For if, for instance, a fly species with very low fossilization potential, due to its poorly preservable habitat and the lack of strongly mineralized skeleton, is found in the fossil record exclusively in, say, Oligocene amber but has also survived until present, it must be regarded as living also during the Miocene and Pliocene; if it went extinct in the Pleistocene, however, it would be considered as contributing to species richness solely in the Oligocene. Yet the closer we are in time to the Recent, the greater is the likelihood that a fossil species will also have an extant representative and therefore, the greater is the proportion of species inferred rather than documented to have existed. Thus, the apparent species richness of geologically young biotas may be disproportionally—as compared to older biotas—inflated by their proximity to the Recent.

The attention devoted by paleontologists to strata of different geological ages also varies, the greatest effort being spent on the most fossiliferous sedimentary rocks. Moreover, the taxonomic philosophy espoused by paleontological systematists varies from fossil group to fossil group—from ammonites to bivalves, for example, or from trilobites to sponges—and consequently the contributions of the two contrasting philosophies, lumping and splitting of species, vary from one geological time interval to another because different organic groups dominated the biotas of different ages. Thus, one may expect data on species richness in some geological periods to include a larger number of spurious species, based solely on minute individual phenotypic variation; whereas data from other time intervals may include a large proportion of species invisible to the viewer because they are hidden within superspecies that in fact encompass a large number of biological species.

All of these biases can be so powerful as to undermine any straightforward empirical analysis of paleontological data intended to estimate the global species richness through geological time. The observed pattern of species richness in the Phanerozoic, reflecting the sheer numbers of fossil species described by paleontologists from sedimentary strata representative of the various time intervals, is well known in its most general form since the nineteenth century. It envisages the organic diversity increasing substantially before the Devonian, then fluctuating in the Middle Paleozoic but declining in the Carboniferous and especially Permian, and rising again, at first slowly in the Mesozoic but indeed dramatically in the Cenozoic. On the whole, it is a pattern of considerable increase in diversity—a pattern of biotic diversification in the Phanerozoic.

Given all the sampling biases, however, which inevitably distort the paleontological data on global species richness in geological time, the observed pattern of diversity obviously cannot be taken on its face value and regarded as a direct and reliable record of the actual megaevolutionary pattern. Analysis of these biases led David Raup (1972) to assert provocatively that the observed pattern might actually reflect only a biased sampling of a pattern that had achieved its maximum diversity as early as in the Mid-Paleozoic and then maintained an approximate equilibrium diversity at only half the maximum level throughout the remainder of the history of marine animal life on the Earth. With such a wide gulf separating the observed pattern and the conceivable actual one, several attempts have been made—and along very diverse lines of reasoning—to correct for the sampling biases and thus to reconstruct the actual pattern of global species richness through time. Since the quantitative significance of particular biases is unknown, however, it is no wonder that different methods have led to widely different results.

I have already mentioned the classic, nineteenth century view on the history of life's diversity on the Earth; it had been derived by John Phillips (1860) from a rather naive, literal reading of the fossil record. Two remarkably similar, though in one important aspect substantially different, interpretations of the actual pattern of biotic diversification were proposed in the 1970s, based on new data and quite sophisticated arguments. James Valentine (1970) compiled data on the average genus richness of fossil families of marine invertebrate animals in the Phanerozoic and discovered a very dramatic increase in genus to family ratio in the Cenozoic. Assuming that a similar change must also have taken place in the average species richness of genera, he concluded that the observed pattern of diversification has to be corrected by including a much sharper than indicated by the classic interpretation, perhaps five- to even tenfold, increase in global species richness during the last 60 million years or so. Compared to this increase, the observed fluctuations in organic diversity during the Paleozoic and Mesozoic appear to be very minor, but they are represented very much in the same way as in the classic view.

In a later article, Valentine and coworkers (1978) considered the changes in the number of marine bioprovinces through time, as inferred from paleobiogeographic studies. The degree of provinciality of marine life seems to never have been as high as it is presently, and it has been controlled by the changing geographic distribution of continental land masses, patterns of oceanic circulation, and global climatic gradients. By estimating, on the basis of modern marine biota, the average species numbers contained within a high- and a low-diversity bioprovince, Valentine and his colleagues were then able to reconstruct the pattern of global species richness as controlled by biogeographic structure of the marine realm. The resulting estimate of the actual pattern of biotic diversification very closely fits the one that Valentine arrived at earlier.

Raup's (1972) proposal that the biases of the fossil record might be so severe as to allow for the actual global species richness being higher in the Mid-Paleozoic than afterward in the history of the biosphere seems to have been primarily intended as a cautionary note. It assumes unrealistically heavy biases and hence is too extreme to be understood as a serious attempt at the estimate of the actual pattern of diversification. Later, however, Raup coauthored a paper by Gould and others (1977) where an argument was developed for global species richness being maintained at roughly the same level since the Ordovician. This argument was based on a comparison between diversity histories of various organic groups as inferred from the fossil record and diversities of high-rank taxa created artificially by computer simulations. Gould, Raup, and their colleagues observed that for as long as their simulated evolutionary universe was being filled with new taxa, these taxa—although created and changing in diversity at random—generally had their maximum diversities early on in their history. As soon as the evolutionary universe as a whole achieved equilibrium, however, in that a dynamic balance developed between appearances and disappearances of lineages within taxa, the maximum diversities of these artificial taxa were always near the midpoint of their durations. The latter is generally the case with post-Ordovician organic groups, whereas the fossil taxa that appeared in the Cambrian or Ordovician usually have their maximum diversities before the midpoint of their temporal ranges. On this basis, then, Gould and coworkers concluded that the marine realm should have rapidly increased its diversity in the Early Paleozoic and been at equilibrium diversity since the Ordovician. Given the unknown contribution of sampling biases, the observed pattern of diversification might be regarded as compatible with this interpretation. Moreover, the importance of these sampling biases should generally decrease as we move up the taxonomic scale. The equilibrium view of biotic diversification in the world ocean was therefore supported by Sepkoski (1978), who compiled the stratigraphic ranges of marine animal orders, found the pattern of their diversity through geological time to largely agree with Gould et al.'s contention, and explicitly argued that the trends in global species and order richness should be concordant with each other.

Richard Bambach (1977), in turn, focused on species richness of individual communities in marine benthic habitats and observed that although the average species number per community in the stressed nearshore environment has remained roughly constant throughout the Phanerozoic, it approximately doubled around the end of the Mesozoic in more open marine environments. He thus concluded that the global species richness had been fluctuating around an equilibrium since the Mid-Paleozoic until the Late Mesozoic, but it then went up by a factor of two or so. Bambach's method is indeed ingenious in that it seems to circumvent the time-dependent biases of the fossil record, which inevitably plague more direct attempts to estimate global trends in species richness, but it also has two major disadvantages. First, it assumes that the number of various kinds of marine benthic communities—community types, as I suggest to call them (Hoffman 1979)—has been roughly constant through time, while it most certainly has not been, if only because the levels of biogeographic provinciality widely varied in the Phanerozoic. Second, it assumes that the sample of communities Bambach analyzed is at least approximately representative of the marine biota in each time interval considered, while there is no way to ensure this to be the case. In fact, Bambach's sample of Paleozoic communities seems to be strongly biased toward species-poor, or at best moderately diverse, brachiopod- and mollusk-dominated communities at the expense of extremely diverse bryozoan- and echinoderm-dominated communities, which are not at all represented in the sample but whose average species richness is by an order of magnitude higher than the median obtained for their time-equivalents. This estimate of the actual pattern might then be compatible with Valentine's reconstruction if the provinciality were taken into account, but it might also very closely resemble the equilibrium view of global species richness if a more balanced sample of Paleozoic communities were considered.

Given these widely different estimates of the actual megaevolutionary pattern of species richness of the biosphere, and given also the absence of unequivocal quantitative criteria to choose between them, the problem of biotic diversification in the Phanerozoic of course could not be successfully addressed. Any potential explanation for the phenomenon—be it specifically megaevolutionary or based on historical contingencies of the process—could not be tested against empirical data, simply because there was no agreement as to what kind of a pattern should be accepted as the empirical pattern to be explained.

In 1981, however, Sepkoski, Bambach, Raup, and Valentine joined forces to coauthor an article that has since been regarded as the consensus estimate of the actual pattern of changes in marine animal diversity through geological time. They analyzed five different paleontological data sets—the number of trace fossil species observed in marine sequences of variable geological ages, the number of invertebrate species per million years as derived from a sample of the data published in the *Zoological Record,* Bambach's data on species richness of benthic fossil assemblages of different ages, the number of genera and subgenera in particular geological periods as derived from the *Treatise on Invertebrate Paleontology,* and Sepkoski's data on the number of marine metazoan families in particular geological stages of the Phanerozoic—and the intercorrelation between these five patterns was very high, indicating their strong overall similarity.

In principle, this might be a spurious correlation because each of these diversity patterns is correlated to geological time, and their independent correlations to that single variable might produce the appearance of the significant intercorrelation. For example, the statistically significant correlation between stork density and local birthrate in Poland is of course due to the independent negative correlations of both these variables to the degree of urbanization, rather than to babies being indeed brought to their parents by storks. Even when the effects of the correlation of each paleontological data set to geological time was removed, however, the residual patterns have remained largely similar and their intercorrelation maintained its statistical significance. This result suggests that these paleontological data bases do not strikingly deviate from the actual pattern of organic diversity, of which they all are samples. But if so, then either data set can be employed as an estimate of the actual pattern; and Sepkoski's family-level data have soon become a sort of the standard for studies

on biotic diversification, perhaps because the level of their stratigraphic distribution—generally, the geological stage—is the finest.

The consensus estimate of the actual pattern of biotic diversification clearly represents a compromise between the earlier, mutually
incompatible interpretations. On the one hand, Raup and Sepkoski
accepted the view of organic diversity generally increasing rather than
being at equilibrium in the Phanerozoic. On the other hand, Valentine agreed to the increase being much less dramatic than he had
previously envisaged; in Sepkoski's data, taxonomic diversity is not
even twice as high in the latest Cenozoic as it is in the Mid-Paleozoic.
This consensus does not, however, solve all the problems with estimating the pattern of diversification. First, even if Sepkoski et al.'s result
points to a single empirical pattern underlaying the five analyzed data
sets, it does not identify the data set that provides the best estimate of
the pattern. Yet they differ in quite a number of important aspects:
whether there was a Paleozoic plateau in diversity, maintained over
tens if not hundreds of millions of years; whether the global diversity
has been increasing continually, or rather episodically, in relatively
short spurts; when did the post-Paleozoic rise in diversity begin, and
how sharp has it been; and so on. In spite of the consensus, these
questions remain unanswered. Second, the intercorrelation of these
five paleontological data sets has been established by analysis at a
very coarse time scale, at the level of geological periods. Therefore, it
cannot validate statements about the actual pattern at any finer scale
of temporal resolution. The variation in diversity between stages representing jointly a single geological period may or may not reflect
sampling biases rather than the actual pattern, and the result obtained by Sepkoski and his colleagues does not resolve this issue.
Third, Philip Signor (1985) has shown in his analysis of sampling
biases that the margin of uncertainty about the scale of the Cenozoic
increase in species richness may approach an order of magnitude,
thus allowing for the consensus as well as for Valentine's previous
estimate of the pattern of diversification. Signor has moreover shown
that, given the sampling biases, different patterns are to be expected
at different taxonomic levels. In fact, the discrepancy between the
observed order- and lower-level diversity patterns through time is
indeed very dramatic; yet there is no reason to suppose that the step
between the order and the family ranks in the taxonomic hierarchy is
more fundamental than the one between the family and the genus or

species. Thus, a discrepancy can also be expected to occur between the actual patterns of species- and family-richness through time.

Consequently, the summary curve of Sepkoski's compendium of marine animal family diversities in Phanerozoic geological stages can only be employed as proxy for the empirical pattern of biotic diversification, and the extent to which it is biased by the properties of the fossil record and the methods of its investigation can hardly be estimated. This curve is very widely regarded as the megaevolutionary phenomenon of diversification to be explained. In fact, it is sometimes explained in terms of specifically megaevolutionary processes—supplementary to those envisaged by the neodarwinian paradigm—and these attempts constitute a paleobiological challenge to neodarwinism. In order to evaluate them, these megaevolutionary explanations must be tested against what is considered as the empirical pattern and they must be compared to rival explanations. While doing so, however, it is extremely important to keep in mind how inevitably poor and biased are our empirical data on this aspect of the history of the biosphere.

12.2 The Pattern to Be Explained

Given a sober assessment of the amount of noise and bias inherent in the empirical pattern of biotic diversification, there can be little doubt that any rigorous testing of potential explanations is impossible—simply because it might boil down to comparing details of the fits of various theoretical models to noise and bias rather than to the historical biological reality. Only a crude quantitative analysis can be reliable under these circumstances. This frustrating statement is additionally justified by the still unsatisfactory precision of absolute time calibration of the stratigraphic boundaries that constitute the geological time scale. Such calibration is prerequisite to quantitative testing of any explanation for the shape of the curve of biotic diversification in the Phanerozoic. The margin of uncertainty about distances in time that separate particular points in this curve—boundaries between the geological stages, if Sepkoski's data are accepted for testing—is often huge, however. In many instances, especially in the Paleozoic and Mesozoic, it considerably exceeds the commonly assumed durations of these time intervals. A change in the estimate of the length of these intervals would not affect our view of the levels of diversity achieved in various moments in the history of the biosphere, but it could have a dramatic effect on the picture of the rates of diversity increases and decreases, or the tempo of diversification.

One might then argue—and many paleontologists in fact do—that because of all the inherent shortcomings of the data base, any research aimed at analysis of the megaevolutionary phenomenon of biotic diversification in the Phanerozoic is, and must be, a purely academic exercise, very far removed from the real world. I believe, however, that this argument is clearly insufficient to deny value to such research. For it is far from obvious what particular constraints the empirical limitations of fossil diversity data impose on their evolutionary interpretation, and how, if at all, these limitations systematically distort the pattern of diversification. It is conceivable that the significance of the sampling biases is in fact much smaller than sometimes assumed, or that they cancel each other out and produce only large amounts of statistical noise. The problem of the mechanisms of, and the controls upon, biotic diversification is out there as a challenge to our science, regardless of whether the empirical data we can put our hands on are satisfactorily good and reliable. It is up to us to seek the appropriate ways to analyze these data and thus to deal with the problem. And, unquestionably, all those who boldly undertook efforts to this end should be greatly credited for drawing our attention to an important aspect of evolution that might otherwise be neglected.

I therefore discuss in this chapter a number of rival explanations for the pattern of biotic diversification in the Phanerozoic. I focus on the pattern of changes in global diversity of marine animals because the fossil record is so much better for marine sedimentary environments and their inhabitants, as a whole, than it is for any other kind of natural environments on the Earth. And I follow the tradition and employ Sepkoski's family-level data set as the estimate of the empirical pattern of diversity, though I believe that I could with equal validity make use of another estimate, for example, Raup's (1978) tabulation of the numbers of fossil genera and subgenera per time interval described in the *Treatise on Invertebrate Paleontology*.

In spite of the much poorer record of sediments deposited on land, such patterns of diversity through time can, and have been, established for land animals (Benton 1985, Sepkoski and Hulver 1985, Padian and Clemens 1985) and plants (Niklas et al. 1980, 1985), and they can also function as proxy for the actual historical patterns. I conjecture that problems very similar to those I address in the context of marine animal diversification also arise with respect to the other two patterns of taxonomic diversification. I see no logical reason to suppose, however, that solutions to these problems should also be necessarily similar. Hence, I deliberately and explicitly restrict the scope of my discussion here to changes in diversity of marine animals.

Sepkoski's curve of marine animal diversity through time represents summation in each consecutive time interval—a geological stage or a group of stages—of all families actually recorded as well as those inferred to have been present because they are recorded in a previous and a later time interval. Families, however, are members of higher taxonomic units, classes. By assigning each family to an appropriate class, one is then able to obtain from Sepkoski's original data set the history of family diversity within each individual class of marine animals. Each class begins as a single family, sooner or later reaches its maximum diversity, and then declines down to ultimate extinction— that is, unless it recovers, sometimes repeatedly, for a shorter or longer time before extinction, and unless of course it still is extant, in which case one cannot tell whether it actually passed its peak. In the simplest and perhaps also most common case, the history of family diversity of a class presents a kind of a spindle, and this is indeed how the graphic representation of such histories has come to be called— spindle diagrams. The history of the marine fauna of the world can therefore be portrayed as a number of such spindle diagrams of family diversity within classes, each of them having its own unique shape and location in geological time. The summation of these spindle diagrams, approximately 90 of them, over time gives the curve of total family diversity of marine animals.

Sepkoski (1981) undertook to search for an order in this apparently bewildering array of spindle diagrams scattered along the axis of geological time. He conducted factor analysis, which is a multivariate statistical technique capable of finding the commonalities between objects described by the same set of measurable characteristics or between variables measured in the same set of points. By the use of factor analysis on his 90 or so spindle diagrams, Sepkoski found surprising simplicity in the pattern. Three kinds of the spindle diagram are so overwhelmingly common in the history of marine fauna that they account for over 90% of the total variation in spindle shape and temporal location. Sepkoski thus subdivided the total marine fauna he considered into three "great evolutionary faunas," as he called them. These are large groups of marine animal classes represented by similar and similarly located in time spindle diagrams, that is to say, groups of classes sharing similar historical patterns of change in their family diversity through geological time. For instance, bivalves, gastropods, echinoids, and teleost fish all increased dramatically in diversity only after the Paleozoic and still predominate among fossilizable inhabitants of the present-day ocean; hence, they are grouped together, along with a number of smaller, less diverse classes, as the Modern Fauna. Articulate brachiopods, rugose corals, cephalopods, crinoids, and a variety of other groups played the most important role in post-Cambrian Paleozoic seas, and they constitute jointly the Paleozoic Fauna. Trilobites, in turn, had their maximum diversity as early as in the Cambrian, and they represent, along with inarticulate brachiopods and a few minor classes, the Cambrian Fauna. Each of these three evolutionary faunas has its own distinctive evolutionary history in terms of family diversity, and the total diversity of marine fauna at each point in time is the sum total of these three components. Their succession in geological time reflects the historical changes in taxonomic composition of the marine biotas. It is the pattern of these three evolutionary faunas, with their characteristic family diversities and the resulting total diversity of the system, that should be adequately accounted for by any potential explanation of biotic diversification in the Phanerozoic.

12.3   On the Multiphase Logistic Model

To account for this complex pattern—and thus to describe and, eventually, explain it—Sepkoski (1978, 1979, 1984) proposed a simple model based on extrapolation from the ecological theory of island biogeography. The main assumption of this model is that the probabilistic (per family) rates of family origination and extinction are diversity dependent; that is, the rate of origination decreases and the rate of extinction increases with increasing diversity of the world's marine fauna, or at least the rate of origination increases less rapidly with increasing diversity than does the rate of extinction. This assumption follows from an ecological rationale given by MacArthur and Wilson (1967) for diversification of island biotas and later extended by Rosenzweig (1975) to the scale of continents or other large biogeographic units. In the latter context, this rationale amounts to the argument that the average population size of species is bound to decrease with increasing diversity in any limited physical environment, and hence the likelihood of a species becoming extinct due to chance environmental fluctuations should increase whereas the likelihood of a species being fragmented and thus speciating should decrease. Sepkoski essentially extrapolated this argument from the species to the family level of taxonomic hierarchy and from the continent or bioprovince to the global geographic scale.

Under the assumption of converse, or at least convergent, relationships of family origination and extinction rates to diversity, there must exist a point of intersection of the mathematical functions that describe them. This point determines the evolutionary equilibrium where the number of originating families is balanced by the number of families going extinct. The predicted curve of family diversity through geological time—the pattern of biotic diversification—is then expressed by the classic logistic equation and has a typical sigmoidal shape. It rises exponentially at first, but it then asymptotically approaches the equilibrium.

Obviously, such a simple logistic model does not provide a satisfactory fit to the empirical pattern of biotic diversification in the Phanerozoic. It can only account for a monotonically increasing curve of diversity through time, but not for a curve including very substantial drops in diversity, and it cannot describe the pattern of change in taxonomic composition of the world's marine fauna. Sepkoski therefore developed a model that involves interaction of three logistic equations representing each one evolutionary fauna. In this model, each fauna increases in diversity according to its own logistic function but it also interacts with the other two faunas in such a way that it begins to decline once its equilibrium diversity is exceeded by the total diversity of the system. This interaction represents a sort of competition between the successive evolutionary faunas. This more complex, multiphase logistic model fits the empirical pattern of diversification much better, for it is capable of reproducing the grand succession of the three evolutionary faunas. When the model is further supplemented with extrinsic perturbations, whose effects on particular evolutionary faunas are turned to the actual effects of the five traditionally recognized mass extinctions on marine animals, it indeed offers a very good fit to the empirical pattern.

Thus, the empirical pattern of biotic diversification can be successfully accounted for by a theoretical model, and the model might be accepted as explanation for this megaevolutionary phenomenon. I

view the multiphase logistic model proposed by Sepkoski as a challenge to the neodarwinian paradigm because it portrays the pattern of
diversification as resulting from action of specifically megaevolutionary processes rather than as a product of historical contingencies of
species origination and extinction. The mathematical structure of this
model envisages biological processes operating on the family level of
taxonomic hierarchy, invoking families as unitary entities. It is the
rates of family extinction and origination that respond in this model
to family diversity of the marine fauna, thus presumably reflecting the
constraints imposed on the process of diversification by the upper
limit on the number of adaptive zones on the Earth. The structure of
the model implies furthermore that the evolutionary faunas interact
and compete with each other as ecological entities during their temporary coexistence. The apparent success of Sepkoski's model may thus
suggest that the neodarwinian paradigm of evolution is indeed incomplete and in need of substantial expansion, since a megaevolutionary
theory is called for to explain the megaevolutionary phenomenon of
changes in global taxonomic diversity.

This corollary is conceptually plausible but before it is accepted,
validity of the megaevolutionary model as the explanation for the
empirical pattern of biotic diversification should be firmly established. The apparent fit between Sepkoski's model and the empirical
pattern, however, tells only that the multiphase logistic model is capable of explaining the pattern but not that it really provides the explanation. One trouble with this model is that its parameters are not, and
cannot be, independently estimated because the empirical pattern is
unique. We have only one biosphere and its one history to investi-
gate. There is no way first to estimate from a separate data set the
rates of family origination and extinction and their dependence on
family diversity in particular evolutionary faunas, and only then to
test whether the curve yielded by the model with these parameters
can reproduce the empirical pattern. The only data source that can be
employed to both these ends is Sepkoski's original compendium of
marine animal families. Of course this problem will also plague all
rival models that may be proposed to account for and explain the
pattern of biotic diversification in the Phanerozoic.

Another trouble with Sepkoski's model, however, is specific to it.
The first step to answer the question whether a theoretical model that
offers a satisfactory fit to any empirical pattern does indeed provide
explanation for this phenomenon must always be to test the model's

assumptions. For if the assumptions are not met, the meaning of the fit between the model and the pattern is unclear. The main assumption of the multiphase logistic model is the one of diversity dependence of the probabilistic (per family) rates of family origination and extinction which leads the system to a dynamic equilibrium diversity on the global scale. Whether this assumption is actually met, however, is a matter of dispute—on both theoretical and empirical grounds.

There must certainly be an upper limit to the number of organic species that can coexist on the Earth. The simplest argument to this effect is that the amounts of energy available for organisms to live on are limited, and hence the number of organisms living at any time must also be limited; consequently, the number of species cannot grow to infinity either, because the number of species obviously cannot be greater than the number of individual organisms. Given the commonness of biparental reproduction, the number of species must in fact be smaller than the number of organisms—half the number at most, but substantially less in reality, because members of exceedingly rare species will have to cope with immense trouble in finding a mate. Thus, there must be an upper limit on species diversity, which implies, in turn, the existence of an upper limit on family diversity, too; the number of families obviously cannot be greater than the number of species, and it must in fact be considerably smaller, or the term "family" would become entirely meaningless.

This conclusion, however, does not in any way imply that the rates of family origination and extinction, or for that matter of speciation and species extinction, have to be diversity dependent. The biosphere may have always been, throughout its long geological history, very far removed from the maximum diversity the Earth can sustain. The eminent plant ecologist Whittaker (1977) argued this point very forcefully and noted also that the appearance of new species creates new resources and thus produces new ecological opportunities for the appearance of further species. The biosphere may also have been kept far from the limit on diversity by physical events, which disturb it and drive smaller or larger numbers of species to extinction before the equilibrium diversity can be reached.

The concept of diversity dependence in origination and extinction of taxa, and hence of the process of diversification, is analogous to the ecological concept of density dependence of birth and death rates in the regulation of population size. The classic ecological theory predicts that the rate of population growth—which is the net result of birth and death rates—should decrease with population density, and hence the pattern of population growth in time should be described by the logistic function and follow a sigmoidal trajectory toward equilibrium. This should be so because the intensity of intraspecific competition between the population members will be at its maximum at very high density and at its minimum in the neighborhood of zero density. Donald Strong (1985) observes, however, that although this may well be true under extremely high and extremely low densities, it is very unlikely that a minor difference in population density can affect the rate of population growth. Empirical data from natural populations document that the population size can fluctuate rather freely under intermediate densities, instead of being strictly and strongly dependent on density. Strong proposes therefore that biological controls on population growth generally are density vague, that is to say, very weak (if they exist at all) under a broad spectrum of ecological circumstances, although quite decisive in extreme situations. In his view, populations can vary tremendously in size and density without these changes being controlled by intraspecific competition, but competition begins to play a significant role when the population has almost reached the maximum size sustainable in a given habitat.

The concept of density vagueness may or may not be adequate as explanation for ecological phenomena concerning the regulation of population growth in nature. It is not the only alternative to the classic concept of density dependence of birth and death rates in the population. Adam Łomnicki (1987), for example, provides an alternative explanation in his provocatively titled book, *Population Ecology of Individuals;* he derives it from individual variation among organisms within populations. But regardless of the ecological validity of density vagueness, it can serve as a very instructive analogy to what may be the case with biotic diversification. I suggest that the process of diversification, at least when considered on the global scale, may well be regulated by diversity vagueness, with strong diversity dependence in the rates of origination and extinction of taxa near the upper limit on organic diversity, and perhaps also under extremely low diversities—as in the earliest Triassic, just after the greatest mass extinction—but entirely free of diversity dependence during the vast majority of the actual history of the biosphere. The potential existence of equilibrium diversity would then be absolutely irrelevant to what is recorded in the historical pattern of biotic diversification in the Phanerozoic.

The regulation of diversity of insular biotas occurs on an intermediate scale between short-term fluctuations of population size and biotic diversification on the geological time scale. MacArthur and Wilson's theory of island biogeography, which has served as a conceptual springboard for Sepkoski's model of diversification, refers to this intermediate time scale. It predicts that the rate of immigration of new species from the mainland to an island decreases with increasing number of species present on the island, whereas the rate of extinction of local populations increases. The former relationship should be mainly due to declining likelihood that a newcomer to the island will belong to a species still absent from the insular biota; the latter should be primarily controlled by increasing competition for limited resources. This theory, however, has not gone unchallenged in ecology. On the one hand, there is an extremely hot debate about whether the patterns in species richness of various biotas this theory is aimed to explain can also be satisfactorily explained as resulting solely from chance colonization (Strong et al. 1984). On the other hand, the observed variation in immigration and extinction rates of species on islands is often so wide that even though the general pattern seems to be concordant with the prediction of diversity dependence, the concept of diversity vagueness may be more applicable on this scale (for example, Strong and Rey 1982).

We know nothing certain about the controls on the rates of species origination and extinction on the global scale. In theory, one may perfectly well imagine—in accordance with our biological knowledge—a world with the rate of speciation, but not species extinction, controlled by species packing in the biosphere and hence dependent on diversity. One may also imagine a world with the rate of species extinction but not origination, or with both or with none of them, being controlled by global diversity. On purely theoretical grounds, either picture can be argued for with some plausibility, and all of them have in fact been argued (e.g., Rosenzweig 1975, Whittaker 1977, Valentine 1980, Hoffman 1981). But even the assumption that the rate of speciation is indeed diversity dependent does not necessarily imply diversity dependence of the rate of origination of species sufficiently different from their ancestors to be recognized by taxonomists as giving rise to new families. It does not imply, in other words, that the rate of family origination is diversity dependent. Thus, there is no compelling theoretical reason to accept the fundamental assumption of Sepkoski's multiphase logistic model of biotic diversification.

There is no empirical reason to accept this assumption either. As calculated independently by Leigh Van Valen (1985; Van Valen and Maiorana 1985) and myself (Hoffman 1986) from Sepkoski's compendium on marine fauna in the Phanerozoic, the probabilistic (per family) rates of family origination and extinction decrease, whereas the family diversity increases with geological time. Given the highly significant correlations of these three megaevolutionary variables to time, spurious correlations might easily arise between them. Nevertheless, there is no significant intercorrelation between either extinction or origination rate and diversity; the rates seem to be diversity independent (Hoffman 1986). Thus, the assumption of the multiphase logistic model of biotic diversification is not met by the empirical pattern of diversity. The crude quantitative analysis I performed actually shows only the absence of a linear or approximately linear dependence of the rates of family origination and extinction on family diversity. Some more complex relationship might in principle exist and validate the logistic model. There is absolutely no evidence for such a relationship, however, and to assume it would certainly be a purely ad hoc solution, accepted solely to defend the model. It seems therefore that Sepkoski's logistic model of biotic diversification is at this point unjustified—and the meaning of its fit to the empirical pattern questionable—though of course the model may still be validated by further research.

12.4 Nonequilibrium Models

The absence of evidence for diversity dependence of the probabilistic (per family) rates of family origination and extinction also undermines the model of diversification proposed recently by Jennifer Kitchell and Timothy Carr (1985), which is also based on this fundamental assumption of diversity dependence. In this case, however, this assumption is not derived by extrapolation from the theory of island biogeography but it is accepted without any further biological justification. Kitchell and Carr's model describes the biosphere as a nonequilibrium system, with mass extinctions and major evolutionary innovations disturbing it and preventing it from ever reaching the equilibrium. The concept of global evolutionary equilibrium—

represented in the model by the upper limit on diversity, which remains constant until a novel evolutionary invention shifts it further upward—is only a theoretical abstract toward which the system evolves between perturbations. The assumption of diversity dependence on the family level of taxonomic hierarchy and on the global scale makes this model megaevolutionary in the same sense as it is the case with Sepkoski's model of biotic diversification.

Kitchell and Carr's model very nicely fits the empirical pattern of three evolutionary faunas and their total diversity through geological time. Moreover, it does so even though its parameters are estimated from only a small subset of the original data but then employed for testing the model against the entire data set. This procedure at least partly overcomes the main obstacle to testing explanations for a unique historical pattern. The trouble with apparent diversity independence of the rates of family origination and extinction, however, pertains to this model as well as to the multiphase logistic one.

Another nonequilibrium model of biotic diversification is advocated by Joel Cracraft (1985), who suggests that the primary controls on the rates of speciation and species extinction are exerted by a plexus of physical environmental factors—the number of geographic and climatic barriers to migration, global climatic gradients, conspicuousness of topographic relief, and so on—which he jointly calls, "lithospheric complexity." Cracraft views the biosphere as an open thermodynamic system maintained by the influx of energy and matter. As such, it appears to him unlikely to ever reach an equilibrium; the corollary that open thermodynamic systems are nonequilibrium systems has in fact been abundantly argued for on the basis of physical principles. Hence, Cracraft postulates that the global diversity has not been and will never be—at least for as long as life is likely to exist on the Earth—at evolutionary equilibrium, even though it may have apparently remained at an approximate plateau for quite a long time in the Paleozoic. Geological processes cause by their very nature incessant changes in the physical environment of life, and they in fact promote increasing complexity of the physical structure of the Earth's surface by producing ever new mountain ranges and blocks of continental crust which are then accreted to older elements. In Cracraft's view, the biosphere should therefore also be in continual flux, leading to ever increasing taxonomic diversity, although mass extinction events may temporarily reverse the trend.

An obvious and rather unusual feature of this model is that it downplays biological interactions between species as having no significance whatsoever for megaevolution. It thus stands in a dramatic opposition to all models of biotic diversification that assume diversity dependence as the main control on the rates of origination and extinction of taxa. It also clashes with the famous Red Queen hypothesis proposed a dozen years earlier by Leigh Van Valen (1973) as the explanation for the apparently incessant, never-ending evolution of taxa. Van Valen's point is that the complexity of biological interactions inevitably produces an ever imminent danger of extinction for each taxon and hence forces all taxa to evolve, even in the absence of any change in the physical environment. Cracraft's argument now runs in exactly the opposite direction. He perceives a continual evolution on the global scale and proposes that it is the change in the physical environment that is the main driving force of this process.

It is not an easy task to determine which of these two contrasting perspectives on evolution more adequately describes the real evolutionary processes. Both viewpoints represent the extremes of a conceivably continuous spectrum of possibilities. Moreover, both are only verbal models, which have first to be quantified in order to allow for their actual testing. Nils Stenseth and John Maynard Smith (1984) undertook to develop in a more rigorous fashion the opposition between the Red Queen and its alternative, which they called, "the Stationary model." Jennifer Kitchell and I attempted an empirical test of these models of evolution in multispecies systems (Hoffman and Kitchell 1984, Kitchell and Hoffman 1989). Unfortunately, their predictions are clearly discernible only under the most extreme circumstances—in the absence of physical environmental change. No real biological system, however, exists in an ecological vacuum, and hence this precondition to successful testing is not met. We cannot tell at the moment whether biological interactions do indeed play a decisive role in driving evolution to go on. We know of course that they do play a role—the arms races between predators and their prey are sufficient evidence of this—but it remains an open question if this role is important enough to refute Cracraft's claim of the overwhelming preponderance of lithospheric complexity in megaevolution.

Perhaps the main trouble with Cracraft's model of biotic diversification, however, is that the concept of lithospheric complexity is too vague to allow for a meaningful testing. For, if we do not know how to quantify lithospheric complexity of the Earth's surface at any moment in time, how could we then even test whether the actual, historical pattern of change in this parameter correlates with the empirical pattern of diversification? Yet the existence of such a correlation—the fit between the model and the pattern it is intended to describe and explain—is the necessary condition for any purported explanation of this megaevolutionary phenomenon. Until this condition is demonstrably met, Cracraft's model cannot be seriously evaluated, even though it may well come closest to truth about the controls on biotic diversification.

12.5 A Neutral Model

The one feature that all these models, both equilibrium and nonequilibrium, have in common is their explicit intention to explain the pattern of biotic diversification in the Phanerozoic by reference to a single set of general laws. These laws may be purely biological, as in the case of diversity dependent models proposed by Sepkoski and Kitchell and Carr, where the entire orderliness of the process of diversification stems from a property of the biosphere itself, given only its ultimate limitation by the available resources. Or they may link the biosphere to its global physical environment, as in the case of Cracraft's postulate that diversity is primarily controlled by lithospheric complexity. In each case, the general laws are of course understood to operate within the bounds of extrinsic conditions of life on the Earth. This important qualification is expressed, for instance, by the assumption that mass extinction events represent phenomena belonging to a different realm of reality than diversification and hence perturbing the biotic system rather than being an inherent part of the controls upon its development through geological time.

Yet, intuitively, it is far from obvious that the megaevolutionary pattern of biotic diversification in the Phanerozoic should be explained by a single set of general laws. Perhaps no such laws are to be invoked to explain it. It is in fact a consequence of the neodarwinian view of megaevolution that this pattern should be explicable in terms of historical contingencies of the processes of speciation and species extinction—as a net outcome of myriads of microevolutionary processes going on at least partly independently, though controlled in part by species interactions and in part by their reactions to the same events in the physical environment, in hundreds of thousands and even millions of species. Current estimates of species richness of the living biosphere run up to 20 to 30 million species; and although the vast majority of these seem to be tropical land animals, mainly insects, this figure gives a hint about the order of magnitude of the total number of species—and hence the number of microevolutionary processes within populations and species—one is talking about while considering the pattern of biotic diversification throughout the 600 million years of the Phanerozoic. A great number of independent, even strictly deterministic, processes can easily result in a random pattern; their totality can mimic chaos. It appears sensible therefore to compare the empirical pattern of biotic diversification in the Phanerozoic to what can be expected to arise by chance. A neutral model of diversification should be evaluated as an alternative and potential rival to the three models discussed previously.

I have proposed a very simple neutral model assuming only that the average probabilistic (per species) rates of species origination and species extinction in the global marine fauna per geological stage vary at random and independently from each other (Hoffman 1986). Thus, the model envisages the pattern of biotic diversification as resulting from summation of two patterns produced by two independent random walks. The assumption of this double random walk model hardly needs a theoretical jusification because it simply implies that the average global rates of speciation and species extinction in several-million-year-long time intervals are abstract statistical constructs reflecting a myriad of independent biological processes operating at the level of microevolution; it reflects the essential unpredictability of an extremely complex historical process. This fundamental assumption of my neutral model is at least plausible as a first approximation.

This neutral model does not imply anything concerning the mutual relationship between the average rates of speciation and species extinction. The very fact, however, that the biosphere has persisted for so long and achieved the bewildering species richness we encounter today points to an obvious empirical constraint on this model of two random walks. The mean rate of species origination over the entire Phanerozoic must have been higher than the mean rate of species extinction. This is entirely consistent with the common sense, for it is only a single local population of a species that needs to diverge genetically from the others in order to give origin to a new species, whereas all local populations must be exterminated to result in species extinction.

The double random walk model deliberately intends to consider the megaevolutionary phenomenon of biotic diversification in the Phanerozoic in terms of microevolutionary causal processes. It can be tested by comparison between the pattern of diversity it predicts and the empirical pattern. It must be kept in mind, however, that the empirical pattern is at the family level, whereas the predictions of this neutral model concern the species level of taxonomic hierarchy. This discrepancy reflects the unfortunate but inescapable limitations of the fossil record. As a consequence, results of the test of the model must be qualified to account for incongruities between the patterns expected to occur at the family and the species levels. A truly rigorous testing of the model is impossible.

Given this frustrating limitation stemming from unsatisfactory paleontological data, the double random walk model withstands the test of the little data we can employ for the purpose of its evaluation (Hoffman 1986). At the family level, both the patterns of origination and extinction pass the test for random walk. Neither the rates of extinction nor the rates of origination deviate from random distributions, though a couple of data points representing extraordinarily high rates may occur in each case. These results corroborate the model, although it cannot be ruled out that some orderliness of species-level rates is hidden under the apparent randomness of family-level phenomena. The only apparent deviation from randomness seems to be the systematic decline of the rates of family origination and extinction through geological time. This decline, however, occurs at the family, not the species, level of taxonomic hierarchy, while there are at least three major factors that may conspire jointly to produce such a pattern even under stochastic constancy of the average rates at the species level. David Raup (1983) has demonstrated that the hierarchical taxonomic structure imposed to the branching evolutionary tree inevitably results in higher rank taxa reaching their maximum diversity earlier than lower rank taxa. The Phanerozoic family origination rates may therefore decline through time even if speciation rates are constant, simply reflecting an artifact of the taxonomic structure of the biosphere. Karl Flessa and David Jablonski (1985) argue that the Phanerozoic family extinction rates may decline in spite of a constancy in species extinction rates, since the average species richness of marine animal families seems to have substantially increased through geological time, whereas species-rich families are of course less vulnerable to stochastic extinction. Finally, the paleontological bias known as the pull of the Recent tends to decrease the extinction rate of families in progressively younger stages because given the same morphologic gap between fossils and living organisms, a Cenozoic species is more likely than a Paleozoic one to be assigned by the paleontologist to an extant family (Raup 1972). This bias tends to extend the actual stratigraphic ranges of many families backward and hence to decrease also the rate of family origination in progressively younger stages. The actual contributions of all these biasing factors to the observed decline in family origination and extinction rates in the Phanerozoic cannot be quantitatively evaluated, at least at the present state of the knowledge; but they ensure that the pattern of apparently declining rates does not refute my neutral model of diversification.

The double random walk model thus withstands at least a preliminary testing. But is it also capable of reproducing the empirical pattern? This question is particulary relevant because the unsatisfactory paleontological data do not allow for adequate testing of the model's predictions. Gene Fenster and I undertook a simulation approach to see if the family of patterns produced by the double random walk model can yield a pattern comparable to the empirical one (Hoffman and Fenster 1986). Take a branching "evolutionary" tree generated at random on the assumption that each branch, or lineage, from the initial one on, has at each point in time 10% probabilities of either further branching or being terminated and 80% probability of persisting unchanged. A summation of the number of lineages within such a tree at each time step gives its lineage richness, which can be represented in the form of a spindle diagram. Spindle diagrams can then be distributed along the time axis. Summation of all spindle diagrams at a time step gives total diversity of the system at that point in the simulation. We distributed 60 such spindle diagrams at random through 50 time steps. This simple simulation corresponds to the double random walk because the rates of lineage origination (branching) and extinction (termination) are random and independent of each other, while the continual, though random, addition of new spindles to the system ensures that the mean rate of origination exceeds the mean rate of extinction.

Not surprisingly, the resulting pattern of total diversity of the system through time is a simple, monotonically rising curve. With the mean rate of origination being greater than the mean rate of extinction, no other result should be expected. The pattern substantially changed, however, when we introduced to the system 15 additional randomly generated spindles with their beginnings clustered around the fifteenth time step, and when we also terminated 10 randomly chosen spindles at the thirty-third time step. These two big events in the history of the simulated system can be regarded as intervention functions, or randomness of an order different from the chance events that control the branching and termination of individual lineages. They are in fact taken to represent two greatest events in the history of the global marine fauna: the Cambro–Ordovician wave of origination and the Permo–Triassic mass extinction, both of which appear exceptional in the frequency distribution of the rates of origination and extinction, respectively. As a component of the model, they closely resemble mass extinction events in the models proposed by Sepkoski, Kitchell and Carr, and Cracraft. Given this modification of the double random walk model, our simulation very nicely fits the empirical pattern of biotic diversification in the Phanerozoic; and when factor analysis is performed on the simulated pattern, the structure of three great evolutionary faunas is also reproduced (Hoffman and Fenster 1986).

I feel that it is fair to say that the empirical pattern can be accounted for by the double random walk model; the neutral model cannot be rejected, as a statistical null hypothesis, by the available evidence. This may imply that the data are so severely affected by various sampling biases, stratigraphic uncertainty, and arbitrary taxonomic decisions that the statistical noise overwhelms the underlying orderly pattern, if there indeed is any. In other words, the neutral model may be too null to be rejected given the unsatisfactory empirical data. But that the empirical pattern of diversification in the Phanerozoic can be described by the double random walk model may also suggest true randomness of the actual process of biotic diversification. In this case, it would not of course imply that individual speciation and species extinction events are indeterministic. It would only mean that the average rates of species origination and extinction for the global marine fauna are determined at each geological stage by myriads of independent factors, and hence the patterns of their change through geological time conform to randomness. The overall patterns of extinction and origination—and consequently also of biotic diversification—would be random, or better, indistinguishable from randomness. They would have to be considered as the net result of a random series of events, like a series of heads and tails while flipping a coin. Analysis of these general megaevolutionary patterns alone could not therefore justify any inferences about their supposed deterministic causes; it could not lead to discovery of any megaevolutionary laws. Nevertheless, all the individual events that constitute these patterns—that is, mass extinctions and waves of originations of any scale—would still be fully deterministic and worth a detailed causal study.

There is obviously no way to determine which of the two interpretations of the apparent fit between the neutral model and the empirical pattern of diversification—overwhelmingly noisy data or actual randomness of the pattern—is more appropriate. Whichever interpretation is chosen, however, it has a profound implication for the understanding of megaevolution. If the paleontological record of biotic diversification is so obscured by statistical noise that no order can be discerned, this megaevolutionary phenomenon cannot provide evidence for existence of any megaevolutionary laws supplementary to those envisaged by the neodarwinian paradigm of evolution. If, on the other hand, the record can be regarded as suggesting that this megaevolutionary pattern reflects only randomness arising at the intersection of a vast number of partly independent microevolutionary processes, there is no need to invoke any megaevolutionary laws as its explanation.

It is important to realize, however, that although our simple simulation fits the empirical pattern of biotic diversification and my crude quantitative test does not reject the double random walk nature of the pattern, this result does not demonstrate that the neutral model is correct but only that there is at present no reason to reject it. It provides a viable explanation for the empirical pattern. None of the other three models, however, has been ultimately rejected either. Given all the limitations of the fossil record, all that can be safely said about the assumption of diversity dependence of origination and extinction rates is that it lacks empirical support but not that it is decisively refuted. Thus, the models proposed by Sepkoski and Kitchell and Carr can be defended. Cracraft's model, in turn, can hardly be subject to any quantitative test at the moment, but this is not a proof of its incorrectness. We thus are on the horns of a real dilemma: How do we decide which of the rival models should be accepted, at least provisionally, as the explanation for the empirical pattern?

The criteria that are to be used in support of the choice of one or another explanation must then be methodological (Hoffman 1987).

Moreover, the unsatisfactory nature of the data must be kept in mind, since it would undermine any attempt to apply a statistical comparison between the fits to the empirical pattern offered by the competing models. Such an attempt might amount to measuring the fit of the models to statistical noise rather than to the record of a real historical biological phenomenon. While facing this dilemma, my own preference is for the double random walk simply because it does not require introduction of a set of evolutionary laws in addition to the laws of evolution within populations and species. Its apparent, and perhaps only provisional, success suggests that there is at present no need for a megaevolutionary theory of biotic diversification. My choice of this model stems of course from my choice of pragmatic reductionism as the methodological criterion to be employed in such situations. Pragmatic reductionism demands that no new levels of causal explanation be introduced as long as there is no demonstrated need to do so, that is, until some empirical observations cannot be satisfactorily accounted for by the laws we know to operate in nature. Since we know that microevolutionary processes occur now and must have occurred in the Phanerozoic, and since we know that they are now and must have been in the geological past conditioned by a variety of historical contingencies, the explanation that refers solely to these components has in my view a methodological priority when compared to megaevolutionary laws, for which there is no independent evidence.

Not everyone is likely to agree with my choice. But whoever prefers to advocate another model of diversification should also spell out clearly the methodological criteria he or she accepts as guidance in solving the dilemma of this megaevolutionary phenomenon.

EPILOGUE

13 The Causal Plexus of Evolution

This book presents one long argument for a certain view of life's history on the Earth. The conclusions to be drawn from my critique of the modern paleobiological challenges to the neodarwinian paradigm are very straightforward. No order has thus far been discovered in the historical patterns of biological evolution that would call for an explanation in terms of specifically macro- or megaevolutionary laws. Such large-scale historical biological patterns as the patterns of phenotypic evolution along particular evolutionary lineages, taxic evolutionary trends within monophyletic groups, mass extinctions, and the pattern of biotic diversification in the Phanerozoic can all be represented as a summation of vast amounts of more or less independent smaller-scale phenomena which find adequate explanations within the neodarwinian paradigm. This is why they are satisfactorily described by stochastic models. They are thus entirely compatible with the neodarwinian assertion that all evolution is an outcome of the interaction between microevolutionary processes operating within myriads of individual populations and species and the global environment evolving over millions of years. Given my explicit methodological choice of what I call pragmatic reductionism, historical biological patterns should be viewed in this way. The neodarwinian paradigm provides the best currently available explanation for macro- and megaevolutionary patterns. From a paleontologist's perspective, therefore, there is at present no reason to regard neodarwinism as either flawed, or at least incomplete.

The force of this conclusion obviously hinges upon the qualifications indicated by my choice of the methodological criteria to rely on, the limits of the field of scientific enquiry I have chosen to discuss, and the moment I have happened to end writing this book. Had I decided to rely solely on the statistical likelihood criterion in evaluating scientific explanations, or to learn enough of molecular biology to be able to make a judgment on various hypotheses in genetics, or to postpone my writing for 10 more years until something new happens in paleobiology, my conclusion might be very different. This conclusion depends, however, at least equally crucially on my understanding of the neodarwinian paradigm.

The assertion that all evolution results from interaction between microevolutionary processes and environment puts equal emphasis on three necessary components of any neodarwinian evolutionary explanation for a historical biological phenomenon: the genetical evolutionary forces, the biological and ecological context of their action within each individual population or species, and the environment. None of these components must be neglected; though one or another may be deliberately, for the sake of simplicity, omitted from a model put forth as a provisional explanation.

Stephen Jay Gould (1977) depicts evolutionary biology as eternally torn between two extreme interpretations of the main controls on evolution. One view emphasizes the intrinsic, purely biological agents, the other points to the extrinsic environment as the prime mover. The neodarwinian paradigm, however, does not fit this conceptual framework because it does not conceive of either one, or the other category of factors as prevailing in evolution. For neodarwinism, as I understand it, both are equally crucial. Moreover, they can bring about evolution exclusively in a mutual interaction mediated by the existence of living beings. The living beings codetermine and influence their environment, which always includes both physical–chemical and biological components, while their structure and function constrain the genetical evolutionary forces, which always operate upon them. Their nature belongs therefore to the causal plexus of all evolution. This is why evolution is inevitably a Markovian process, where the current state of a biological system or entity has an impact on its future state. Consequently, evolution is inherently historical in nature—which it would not be if it were to result solely from the interplay between evolutionary forces and environment—and this fact must be taken into account while undertaking to seek an evolutionary explanation for any biological phenomenon.

This complexity of the causation of revolutionary phenomena is rather obvious, though perhaps not sufficiently emphasized by neo-darwinians, for microevolution (Lewontin 1983). It should be at least equally obvious for macro- and megaevolution. These large-scale historical biological patterns must not be considered in isolation from what has been going on in the environment. The evolution of environment on the Earth, moreover, must always be analyzed jointly with the evolution of the biosphere. These statements may sound trivial, but their consequences can be far-reaching.

Consider, for example, the evolution of calcareous skeletal structures in marine organisms. For approximately three billion years, organisms lived and evolved in the ocean but they did not produce skeletons. Starting with the Cambrian, however, several groups began to secrete calcium carbonate matter in the form of ever more elaborate skeletal structures. A possible explanation for this megaevolutionary pattern is to view skeletons as primarily defensive devices against predators and to interpret their common appearance in the Cambrian seas as an adaptive response to the origin of the first large predators and the apparent trend toward their increasing thickness and complexity, especially in the late Mesozoic and Cenozoic, as a result of the continually escalating arms race between shell-crushing animals and their prey. Geerat Vermeij (1987) is the most prominent proponent of this perspective on life as a largely autonomous system.

An alternative view stems from the realization of the tremendous role that calcium plays in the functioning of the living cell, especially in the eukaryota, and from a model of the chemical evolution of the world's ocean. All living cells keep the calcium concentration in cytosol at an extremely low level, several orders of magnitude lower than the levels encountered in the modern ocean. They have very elaborate mechanisms to regulate calcium content; and they are very sensitive to both excess and deficiency of this element in the environment, for although the presence of calcium is absolutely critical for many physiological functions, it becomes deleterious and even lethal at higher concentrations. This information on modern organisms suggests then that any changes in the level of calcium concentration in the ocean must have influenced the evolution of life in the past. On the other hand, Stefan Kempe and Egon Degens (1985) argue that the chemistry of the early ocean was very different from the modern one. They envisage an alkaline soda ocean—poor in calcium but rich in sodium and dissolved carbonates—in which the earliest organisms evolved more than three billion years ago. The calcium concentration in this soda ocean could be as low as it is now in the living cell. During the later evolution of the Earth, however, the concentration of calcium in the ocean was building up, with a sort of threshold being reached around the beginning of the Cambrian, when calcium and magnesium replaced sodium as the cations balancing the dissolved carbonates. The rising levels of calcium in the environment presented a challenge to all organisms, and the production of calcareous skeletons may well be viewed as an adaptive solution to this problem—a form of calcium detoxification. Both the onset of skeletal structures in the history of life on the Earth and the subsequent trend toward heavier skeletons of marine organisms, with its episodic ups and downs, appear then primarily as a response to the changes in the physical environment of life (Kaźmierczak et al. 1985).

This evolution of calcareous skeletal structures in marine organisms may in fact be less than the main effect that the increasing concentration of calcium in the ocean has had on the biosphere. Józef Kaźmierczak and Egon Degens (1986) point out that, in living organisms, the capacity of amoebas for endocytosis and the ability of slime mold and sponge cells to aggregate dramatically decrease if the calcium concentration in their environment falls below a certain level, somewhere between the levels normally encountered in the marine environment and those maintained in the cytosol. They propose that the chemical evolution of the early ocean toward increased calcium concentrations thus ensured that the fundamental conditions were met for the origin of the eukaryotic cell through endosymbiosis and for the later development of multicellular organisms.

The hypothesis of changes in calcium concentration in the ocean as a promoter of megaevolutionary changes in the biosphere may sound as if it portrayed life as directed entirely by the abiotic environment—thus representing exactly the opposite view to Vermeij's perspective. Actually, however, these megaevolutionary phenomena do not simply reflect a response of life to the environmental stimulus. It is entirely conceivable that life forms could evolve for which calcium concentrations would be physiologically irrelevant, or which would find ways to adapt to a very wide range of calcium levels in the environment. Yeast may in fact demonstrate that this supposition is not too far-fetched, for they apparently have developed an efficient mechanism to allow for survival even under extreme fluctuations in calcium concentration (Kovač 1985). Perhaps this, or some other presently nonexistent or merely not yet discovered, regulatory mechanism could have spread in the biosphere billions of years ago. In any event, the exceptional calcium physiology of yeast shows that it is history, rather than purely biochemical necessity, that has permitted calcium to play such a role in the evolution as envisaged by Kaźmierczak and Degens. The historically established dependence of the vast majority of the eukaryota on calcium mediates then the interaction between long-term changes in the ocean chemistry, on the one hand, and the genetical evolutionary forces within individual populations and species, on the other.

Certainly, the case for the role of calcium in megaevolution is not compelling, if only because independent evidence for the increasing calcium concentrations in the ocean is at present hardly available. Yet even its plausibility alone demonstrates the need to consider all evolution as a fundamentally historical process, driven by interaction between life and the environment specific to each population and species. Calcium concentration in the environment is but one among myriads of factors that affect each organism and, consequently, the evolutionary history of each organic group. Undoubtedly, predation pressure—which is so strongly emphasized by Vermeij—but also all other biological agents, including cooperative and even symbiotic relationships, exert powerful influences upon all populations and species. Thus, innumerable environmental factors pose challenges and constrain the potential responses to be effected by simultaneous action of a variety of genetical evolutionary forces; the spectrum of these potential responses is further limited by the current state of each organism, population, and species. Every step in evolution of each organic group results from this complex web of interactions which constitutes the causal plexus of evolution. To forget about any of the three components of this plexus—the various genetical evolutionary forces, the multifaceted physical and biological environment, and the nature of each particular group achieved as a result of its evolutionary history—inevitably leads to an oversimplified, and hence distorted, picture of the evolutionary process. This corollary of the neodarwinian paradigm cannot be emphasized too strongly.

The very complexity of the causal plexus of evolution and its ineluctably historical nature—dependent not only on the local and momentary configuration of the environmental conditions and the accidental occurrence of genetical mutations leading under these circumstances to a particular phenotypic effect, but also upon the past evolutionary history of each species—have profound consequences for our understanding of evolution. For it follows that evolutionary explanations of historical biological phenomena must always be phrased within what I have termed the "individualistic," instead of "uniformistic," framework (Hoffman 1981).

The uniformistic approach, where each object or entity is an element of a class of identical objects, is fully justified in physics and chemistry because individual electrons, atoms, and molecules are indeed identifiable by their positions in space and time but not by their unique characteristics. This is not only an axiom but also an empirical statement, if only because of the uncertainty principle. The order of the physical microcosm depends therefore on universal laws rather than on individual particles. Evolutionary biology, in turn, is fundamentally individualistic because each organism, population, and species has its own historically established features which codetermine, along with the universal laws of microevolution, their fate in evolution. The apparent order of the biosphere is therefore a byproduct of the ecological and evolutionary behavior of millions upon millions of individual biological entities. This is why the patterns of megaevolution can be described in stochastic terms—mass extinctions as clusters of independent events, the pattern of biotic diversification as a result of two random walks—for chance arises at the intersection of the evolutionary processes that go on incessantly in vast numbers of individual populations and species. This chance, however, is always constrained by history.

References

Abel, O. 1904. Über das Aussterben der Arten. *IXème Congr. Géol. Int., Vienne,* 739–748.

Alvarez, L. W.; Asaro, F.; Michel, H. V. and Alvarez, W. 1980. Extraterrestrial cause for the Cretaceous-Tertiary extinction. *Science* 208, 1095–1108.

Alvarez, W.; Alvarez, L. W.; Asaro, F. and Michel, H. V. 1982. Iridium anomaly approximately synchronous with terminal Eocene extinction. *Science* 216, 886–888.

Alvarez, W.; Kauffman, E. G.; Surlyk, F.; Alvarez, L. W.; Asaro, F. and Michel, H. V. 1984. Impact theory of mass extinctions and the invertebrate fossil record. *Science* 223, 1135–1141.

Amsterdamski, S. 1983. *Między historią a metodą.* Państwowy Instytut Wydawniczy, Warszawa.

Alberch, P. 1980. Ontogenesis and morphological diversification. *Am. Zool.* 20, 653–667.

Arnold, A. J. and Fristrup, K. 1982. The theory of evolution by natural selection: A hierarchical expansion. *Paleobiology* 8, 113–129.

Ayala, F. J. 1975. Genetic differentiation during the speciation process. *Evol. Biol.* 8, 1–78.

Bambach, R. K. 1977. Species richness in marine benthic habitats through the Phanerozoic. *Paleobiology* 3, 152–167.

Bambach, R. K. 1983. Ecospace utilization and guilds in marine communities through the Phanerozoic. In: Tevesz, M. J. S. and McCall, P. L. (eds.), *Biotic Interactions in Fossil and Recent Benthic Communities.* Plenum Press, New York, pp. 719–746.

Barbieri, M. 1985. *The Semantic Theory of Evolution.* Harwood, Chur, Switzerland.

Barigozzi, C. (ed.) 1982. *Mechanisms of Speciation.* Liss, New York.

Barnard, C. J. 1984. Stasis: A coevolutionary model. *J. Theor. Biol.* 110, 27–34.

Bejer-Petersen, B. 1975. Length of development and survival of *Hylobius abietis* as influenced by silvicultural exposure to sunlight. *Kgl. Vet. Landbok. Arsskr.* 1975, 111–120.

Bell, G. and Maynard Smith, J. 1987. Short-term selection for recombination among mutually antagonistic species. *Nature* 328, 66–68.

Bell, M. A.; Baumgartner, J. F. and Olson, E. C. 1985. Patterns of temporal change in single morphological characters of a Miocene stickleback fish. *Paleobiology* 11, 258–271.

Bell, M. A. and Haglund, T. R. 1982. Fine-scale temporal variation of the Miocene stickleback *Gasterosteus doryssus*. *Paleobiology* 8, 282–292.

Benton, M. J. 1985. Patterns in the diversification of Mesozoic non-marine tetrapods and problems in historical diversity analysis. *Spec. Pap. Palaeont.* 33, 185–202.

Benton, M. J. 1987. The Late Triassic tetrapod extinction events. In: Padian, K. (ed.), *The Beginning of the Age of Dinosaurs*. Cambridge University Press, Cambridge, pp. 303–320.

Bertalanffy, L. von. 1949. *Das Biologische Weltbild*. Franske, Bern.

Bethell, T. 1985. Agnostic evolutionists: The taxonomic case against Darwin. *Harper's Magazine*, February 1985, 49–61.

Blandino, G. 1960. *Problemi e dottrine di biologia teoretica*. Torino.

Blank, R. G. and Ellis, C. H. 1982. The probable range concept applied to the biostratigraphy of marine microfossils. *J. Geol.* 90, 415–434.

Bookstein, F. L.; Gingerich, P. D. and Kluge, A. D. 1978. Hierarchical linear modeling of the tempo and mode in evolution. *Paleobiology* 4, 120–134.

Brenchley, P. J. 1984. Late Ordovician extinctions and their relationship to the Gondwana glaciation. In: Brenchley, P.J. (ed.), *Fossils and Climate*. Wiley, Chichester, pp. 291–316.

Bretsky, P. W. and Klofak, S. M. 1985. Margin to craton expansion of Late Ordovician benthic marine invertebrates. *Science* 227, 1469–1471.

Brinkmann, R. 1929. Statistisch-biostratigraphische Untersuchungen an mitteljurassischen Ammoniten über Artbegriff und Stammesentwicklung. *Abh. Ges. Wiss. Göttingen, Math.-Phys. Kl., N.F.* 13(3), 1–249.

Brocchi, G. 1814. *Conchiliologia Fossile Subapennina*. Milan.

Brooks, D. R. and Wiley, E. O. 1985. *Evolution as Entropy*. University of Chicago Press, Chicago.

Buckland, W. 1823. *Reliquiae Diluvianae*. Murray, London.

Cahn, P. H. 1958. Comparative optic development in *Astyanax mexicanus* and in two of its blind cave derivatives. *Amer. Mus. Nat. Hist. Bull.* 115, 69–112.

Cannon, H. G. 1959. *Lamarck and Modern Genetics*. University of Manchester Press, Manchester.

Carson, H. L. 1982. Speciation as a major reorganization of polygenic balances. In: Barigozzi, C. (ed.), *Mechanisms of Speciation*. Liss, New York, pp. 411–434.

Cavalli-Sforza, L. L. and Bodmer, W. F. 1971. *The Genetics of Human Populations*. Freeman, San Francisco.

Chaline, J. 1987. Arvicolid data (Arvicolidae, Rodentia) and evolutionary concepts. *Evol. Biol.* 21, 237–310.

Chaline, J. and Laurin, B. 1986. Phyletic gradualism in a European Plio-Pleistocene *Mimomys* lineage (Arvicolidae: Rodentia). *Paleobiology* 12, 203–216.

Charlesworth, B.; Lande, R.; and Slatkin, M. 1982. A Neo-Darwinian commentary on macroevolution. *Evolution* 36, 474–498.

Cheetham, A. H. 1986. Tempo of evolution in a Neogene bryozoan: Rates of morphologic change within and across species boundaries. *Paleobiology* 12, 190–202.

Cisne, J. L.; Chandler, G. O.; Rabe, B. D. and Cohen, J. A. 1982. Clinal variation, episodic evolution, and possible parapatric speciation: The trilobite *Flexicalymene senaria* along an Ordovician depth gradient. *Lethaia* 15, 325–341.

Collins, J. P. and Cheek, J. E. 1983. Effect of food and density on development of typical and cannibalistic salamander larvae in *Ambystoma tigrinum nebulosum*. *Amer. Zool.* 23, 77–84.

Connor, E. F. 1986. Time series analysis of the fossil record. In: Raup, D. M. and Sepkoski, J. J. (eds.), *Patterns and Processes in the History of Life*. Springer, Berlin, pp. 119–147.

Conrad, M. 1983. *Adaptability: The Significance of Variability from Molecule to Ecosystem*. Plenum Press, New York.

Cooper, R. A. 1973. Taxonomy and evolution of *Isograptus* Moberg in Australasia. *Palaeontology* 16, 45–115.

Cope, E. D. 1896. *Primary Factors of Organic Evolution*. Open Court, Chicago.

Cope, J. C. W.; Getty, D. A.; Howarth, M. K.; Morton, N. and Torrens, H. S. 1980a. A correlation of Jurassic rocks in the British Isles. Part 1. Introduction and Lower Jurassic. *Geol. Soc. London Spec. Rept.* 14.

Cope, J. C. W.; Duff, K. L.; Parsons, C. F.; Torrens, H. S.; Wimbledon, W. A. and Wright, J. K. 1980b. A correlation of Jurassic rocks in the British Isles. Part 2. Middle and Upper Jurassic, *Geol. Soc. London Spec. Rept.* 15.

Cracraft, J. 1985. Biological diversification and its causes. *Ann. Missouri Bot. Gard.* 72, 794–822.

Crick, F. H. C. 1968. The origin of the genetic code. *J. Mol. Biol.* 38, 367–379.

Cronin, T. M. 1985. Speciation and stasis in marine Ostracoda: Climate modulation of evolution. *Science* 227, 60–63.

Cuénot, L. 1941. *Invention et Finalité en Biologie*. Masson, Paris.

Cuénot, L. 1951. *L'Evolution Biologique*. Masson, Paris.

Cutbill, J. L. and Funnell, B. M. 1967. Computer analysis of the fossil record. In: Harland, W. B. et al. (eds.), *The Fossil Record*. Geol. Soc. London, London, pp. 791–820.

Cuvier, G. 1799. Memoir sur les éspèces d'élephants vivantes et fossiles. *Mem. Acad. Sci. Paris* 2, 1–32.

Cuvier, G. 1825. *Discours sur les Révolutions de la Surface du Globe et sur les Changements qu'Elles Ont Produites dans le Règne Animal*. Dufour et d'Ocagne, Paris.

Danto, A. C. 1965. *Analytical Philosophy of History*. Cambridge University Press, Cambridge.

Darwin, C. 1846. *Geological Observations on South America*. Smith & Elder, London.

Darwin, C. 1859. *The Origin of Species by Means of Natural Selection*. Murray, London.

Davitashvili, L. S. 1966. *Sovremennoye sostayanye evolutsyonnogo uchenya na zapade*. Nauka, Moskva.

Dawkins, R. 1976. *The Selfish Gene*. Clarendon Press, Oxford.

Dawkins, R. 1982. *The Extended Phenotype*. Freeman, San Francisco.

Dawkins, R. 1986. *The Blind Watchmaker*. Longman, Harlow.

Depew, D. J. and Weber, B. H. (eds.) 1985. *Evolution at a Crossroads*. MIT Press, Cambridge, MA.

De Vries, H. 1905. *Species and Varieties: Their Origin by Mutation*. Open Court, Chicago.

Dingus, L. 1984. Effects of stratigraphic completeness on interpretations of extinction rates across the Cretaceous-Tertiary boundary. *Paleobiology* 10, 420–438.

Dobzhansky, T. 1937. *Genetics and the Origin of Species*. Columbia University Press, New York.

Dobzhansky, T. 1970. *Genetics of the Evolutionary Process*. Columbia University Press, New York.

Douglas, M. E. and Avise, C. J. 1982. Speciation rates and morphological divergence in fishes: Tests of gradual versus rectangular modes of evolutionary change. *Evolution* 36, 224–232.

Dover, G. A. 1982. Molecular drive: A cohesive mode of species evolution. *Nature* 292, 111–117.

Dray, W. 1957. *Laws and Explanation in History*. Oxford University Press, Oxford.

Duhem, P. 1954. *The Aim and Structure of Physical Theory*. Princeton University Press, Princeton.

Edwards, A. 1972. *Likelihood*. Cambridge University Press, Cambridge.

Eldredge, N. 1979. Alternative approaches to evolutionary theory. *Carnegie Mus. Nat. Hist. Bull.* 13, 7–19.

Eldredge, N. 1982. Phenomenological levels and evolutionary rates. *Syst. Zool.* 31, 338–347.

Eldredge, N. 1984. Simpson's inverse: Bradytely and the phenomenon of living fossils. In: Eldredge, N. and Stanley, S. M. (eds.), *Living Fossils.* Springer, New York, pp. 272–277.

Eldredge, N. 1985. *Unfinished Synthesis.* Oxford University Press, New York.

Eldredge, N. and Cracraft, J. 1980. *Phylogenetic Patterns and the Evolutionary Process.* Columbia University Press, New York.

Eldredge, N. and Gould, S. J. 1972. Punctuated equilibria: An alternative to phyletic gradualism. In: Schopf, T. J. M. (ed.), *Models in Paleobiology.* Freeman, San Francisco, pp. 82–115.

Estes, R. 1975. Fossil *Xenopus* from the Paleocene of South America and the zoogeography of pipid frogs. *Herpetologica* 31, 263–278.

Fisher, R. A. 1930. *The Genetical Theory of Evolution.* Clarendon Press, Oxford.

Fitch, W. M. and Margoliash, E. 1970. The usefulness of amino acid and nucleotide sequences in evolutionary studies. *Evol. Biol.* 4, 67–109.

Flessa, K. J. and Jablonski, D. 1985. Declining Phanerozoic background extinction rates: Effect of taxonomic structure? *Nature* 313, 216–218.

Ford, E. B. 1944. Studies on the chemistry of pigments in the Lepidoptera, with special reference to their bearing on systematics. *Trans. Roy. Entomol. Soc. London* 94, 201–223.

Futuyma, D. J. 1979. *Evolutionary Biology.* Sinauer, Sunderland, Mass.

Gadamer, H.-G. 1965. *Wahrheit und Methode,* 2nd ed. Mohr, Tübingen.

Ganapathy, R. 1982. Evidence for a major meteorite impact on the Earth 34 million years ago: Implication for Eocene extinctions. *Science* 216, 885–886.

Gartner, S. and Keany, J. 1978. The terminal Cretaceous event: A geologic problem with an oceanographic solution. *Geology* 6, 708–712.

Ghiselin, M. T. 1974. A radical solution to the species problem. *Syst. Zool.* 23, 536–544.

Ghiselin, M. T. 1987. Species concepts, individuality, and objectivity. *Biol. Philos.* 2, 127–144.

Gilinsky, N. L. 1981. Stabilizing species selection in the Archaeogastropoda. *Paleobiology* 7, 316–331.

Gilinsky, N. L. 1986. Species selection as a causal process. *Evol. Biol.* 20, 249–274.

Gingerich, P. D. 1976. Paleontology and phylogeny: Patterns of evolution at the species level in early Tertiary mammals. *Am. J. Sci.* 276, 1–28.

Gingerich, P. D. 1983. Rates of evolution: Effects of time and temporal scaling. *Science* 222, 159–161.

Gingerich, P. D. 1985. Species in the fossil record: Concepts, trends, and transitions. *Paleobiology* 11, 27–41.

Goldschmidt, R. 1940. *The Material Basis of Evolution.* Yale University Press, New Haven.

Golenkin, M. I. 1927. *Pobediteli v Borbe za Sushchestvovanye.* Bot. Inst. Mosk. Gos. Univ., Moskva.

Goodwin, B. C.; Holder, N.; and Wylie, C. C. 1983. *Development and Evolution.* Cambridge University Press, Cambridge.

Gould, S. J. 1977. Eternal metaphors of paleontology. In: Hallam, A. (ed.), *Patterns of Evolution.* Elsevier, Amsterdam, pp. 1–26.

Gould, S. J. 1980. Is a new and general theory of evolution emerging? *Paleobiology* 6, 119–130.

Gould, S. J. 1982a. The meaning of punctuated equilibrium and its role in validating a hierarchical approach to macroevolution. In: Milkman, R. (ed.), *Perspectives on Evolution.* Sinauer, Sunderland, Mass., pp. 83–104.

Gould, S. J. 1982b. Change in developmental timing as a mechanism of macroevolution. In: Bonner, J. T. (ed.), *Evolution and Development.* Springer, Berlin, pp. 333–346.

Gould, S. J. 1982c. Darwinism and the expansion of evolutionary theory. *Science* 216, 380–387.

Gould, S. J. 1983. Irrelevance, submission, and partnership: The changing role of paleontology in Darwin's three centennials and a modest proposal for macroevolution. In: Bendall, D. S. (ed.), *Evolution from Molecules to Men.* Cambridge University Press, Cambridge, pp. 347–366.

Gould, S. J. 1984. Toward the vindication of punctuational change. In: Berggren, W. A. and Van Couvering, J. A. (eds.), *Catastrophes and Earth History.* Princeton University Press, Princeton, pp. 9–34.

Gould, S. J. 1985. The paradox of the first tier: An agenda for paleobiology. *Paleobiology* 11, 2–12.

Gould, S. J. and Eldredge, N. 1977. Punctuated equilibria: The tempo and mode of evolution reconsidered. *Palebiology* 3, 115–151.

Gould, S. J. and Eldredge, N. 1986. Punctuated equilibrium at the third stage. *Syst. Zool.* 35, 143–148.

Gould, S. J.; Raup, D. M.; Sepkoski, J. J.; Schopf, T. J. M. and Simberloff, D. S. 1977. The shape of evolution: A comparison of real and random clades. *Paleobiology* 3, 23–40.

Grabert, B. 1959. Phylogenetische Untersuchungen an *Gaudryina* und *Spiroplectinata* (Foram.) besonders aus dem nordwestdeutschen Apt und Alb. *Abh. Senckenb. Naturforsch. Ges.* 498, 1–71.

Grant, V. 1963. *The Origin of Adaptations.* Columbia University Press, New York.

Grassé, P.-P. 1973. *L'Evolution du Vivant*. Michel, Paris.

Grivell, L. A. 1986. Deciphering divergent codes. *Nature* 324, 109–110.

Grünbaum, A. 1976. Ad hoc auxiliary hypotheses and falsificationism. *Brit. J. Philos. Sci.*, 27.

Gruszczyński, M. and Małkowski, K. 1987. Stable isotopic records of the Kapp Starostin Formation (Permian), Spitsbergen. *Pol. Polar Res.* 8, 201–215.

Hacking, I. 1965. *The Logic of Statistical Inference*. Cambridge University Press, Cambridge.

Haldane, J. B. S. 1932. *The Causes of Evolution*. Longman, London.

Hallam, A. 1981. *Facies Interpretation and the Stratigraphic Record*. Freeman, San Francisco.

Hallam, A. 1986. The Pliensbachian and Toarcian extinction events. *Nature* 319, 765–768.

Hallam, A. 1987. End-Cretaceous mass extinction event: Argument for terrestrial causation. *Science* 238, 1237–1242.

Hamilton, W. D. 1964. The genetical theory of social behavior: I and II. *J. Theor. Biol.* 7, 1–52.

Hansen, T. A. 1980. Influence of larval dispersal and geographic distribution on species longevity in neogastropods. *Paleobiology* 6, 193–207.

Hansen, T. A. 1982. Modes of larval development in early Tertiary neogastropods. *Paleobiology* 8, 367–377.

Hardy, A. C. 1965. *The Living Stream*. Collins, London.

Harland, W. B.; Cox, A. V.; Llewellyn, P. G.; Pickton, C. A. G.; Smith, A. G. and Walters, R. 1982. *A Geologic Time Scale*. Cambridge University Press, Cambridge.

Hays, J. D. and Shackleton, N. J. 1976. Globally synchronous extinction of the radiolarian *Stylatractus universus*. *Geology* 4, 649–652.

Hecht, M. K. and Hoffman, A. 1986. Why not Neodarwinism? A critique of paleobiological challenges. *Oxford Surv. Evol. Biol.* 3, 1–47.

Hempel, C. G. 1966. *The Philosophy of Natural Science*. Prentice-Hall, Englewood Cliffs.

Hempel, C. G. and Oppenheim, P. 1948. Studies in the logic of explanation. *Philos. Sci.* 15, 135–175.

Hennig, E. 1932. *Wege und Wesen der Paläontologie*. Borntraeger, Berlin.

Hilgendorf, F. 1866. *Planorbis multiformis* im Steinheimer Süsswasserkalk. Ein Beispiel von Gestaltveränderung im Laufe der Zeit. Weber, Berlin.

Ho, M.-W. and Saunders, P. T. (eds.) 1984. *Beyond Neo-Darwinism*. Academic Press, London.

Hoffman, A. 1977. Synecology of macrobenthic assemblages of the Korytnica Clays (Middle Miocene; Holy Cross Mountains, Poland). *Acta Geol. Polon.* 27, 227–280.

Hoffman, A. 1979. Community paleoecology as an epiphenomenal science. *Paleobiology* 5, 357–379.

Hoffman, A. 1981. Stochastic versus deterministic approach to paleontology: The question of scaling or metaphysics? *N. Jb. Geol. Paläont., Abh.* 162, 80–96.

Hoffman, A. 1982. Punctuated versus gradual mode of evolution: A reconsideration. *Evol. Biol.* 15, 411–436.

Hoffman, A. 1983. Paleobiology at the crossroads: A critique of some modern paleobiological research programs. In: Grene, M. (ed.), *Dimensions of Darwinism.* Cambridge University Press, Cambridge, pp. 241–271.

Hoffman, A. 1985. Patterns of family extinction depend on definition and geological timescale. *Nature* 315, 359–362.

Hoffman, A. 1986. Neutral model of Phanerozoic diversification: Implications for macroevolution. *N. Jb. Geol. Paläont., Abh.* 172, 219–244.

Hoffman, A. 1987. Neutral model of taxonomic diversification in the Phanerozoic: A methodological discussion. In: Nitecki, M. H. and Hoffman, A. (eds.), *Neutral Models in Biology.* Oxford University Press, New York, pp. 133–146.

Hoffman, A. 1988. Mass extinctions: The view of a skeptic. *J. Geol. Soc. London,* in press.

Hoffman, A. and Fenster, E. J. 1986. Randomness and diversification in the Phanerozoic: A simulation. *Palaeontology* 29, 655–663.

Hoffman, A. and Kitchell, J. A. 1984. Evolution in a pelagic planktic system: A paleobiologic test of models of multispecies evolution. *Paleobiology* 10, 9–33.

Hoffman, A. and Nitecki, M. H. 1985. Reception of the asteroid hypothesis of terminal Cretaceous extinctions. *Geology* 13, 884–887.

Hoffman, A. and Nitecki, M. H. (eds.) 1986. *Problematic Fossil Taxa.* Oxford University Press, New York.

Hoffman, A.; Pisera, A. and Studencki, W. 1978. Reconstruction of a Miocene kelp-associated macrobenthic ecosystem. *Acta Geol. Polon.* 28, 377–387.

Hoffman, A. and Reif, W.-E. 1988. On methodology of the biological sciences: From a biological perspective. *N. Jb. Geol. Paläont., Abh.,* in press.

Hoffman, A. and Szubzda-Studencka, B. 1982. Bivalve species duration and ecologic characteristics in the Badenian (Miocene) marine sandy facies of Poland. *N. Jb. Geol. Paläont., Abh.* 163, 122–135.

Holser, W. T.; Magaritz, M. and Wright, J. 1986. Chemical and isotopic variations in the world ocean during Phanerozoic time. In: Walliser, O. H. (ed.), *Global Bio-Events.* Springer, Berlin, pp. 63–74.

Hsü, K. J. 1987. *The Great Dying.* Harcourt Brace Jovanovich, New York.

Hull, D. L. 1974. *The Philosophy of Biological Sciences.* Prentice-Hall, Englewood Cliffs.

Hull, D. L. 1976. Are species really individuals? *Syst. Zool.* 25, 174–191.

Hull, D. L. 1980. Individuality and selection. *Ann. Rev. Ecol. Syst.* 11, 311–332.

Hut, P.; Alvarez, W.; Elder, W. P.; Hansen, T.; Kauffman, E. G.; Keller, G.; Shoemaker, E. M. and Weissman, P. R. (1987). Comet showers as a cause of mass extinctions. *Nature* 329, 118–126.

Huxley, J. S. 1942. *Evolution: The Modern Synthesis.* Allen & Unwin, London.

Huxley, T. A. 1862. The anniversary address. *Quart. J. Geol. Soc. London* 18, 40–54.

Jablonski, D. 1980. Apparent versus real biotic effects of transgressions and regressions. *Paleobiology* 6, 397–407.

Jablonski, D. 1986a. Background and mass extinctions: The alternation of macroevolutionary regimes. *Science* 231, 129–133.

Jablonski, D. 1986b. Causes and consequences of mass extinctions: A comparative approach. In: Elliott, D. K. (ed.), *Dynamics of Extinction.* Wiley, New York, pp. 183–230.

Jablonski, D. 1986c. Larval ecology and macroevolution in marine invertebrates. *Bull. Mar. Sci.* 39, 565–587.

Jablonski, D. and Flessa, K. W. 1986. The taxonomic structure of shallow-water marine faunas: Implications for Phanerozoic extinctions. *Malacologia* 27, 43–66.

Jacob, F. 1982. *The Possible and the Actual.* University of Washington Press, Seattle.

Janvier, P. 1984. Cladistics: Theory, purpose, and evolutionary implications. In: Pollard, J.W. (ed.), *Evolutionary Theory: Paths into the Future.* Wiley, Chichester, pp. 39–75.

Johnson, D. A. and Nigrini, C. A. 1985. Synchronous and time-transgressive Neogene radiolarian datum levels in the Equatorial Indian and Pacific Oceans. *Mar. Micropaleont.* 9, 489–523.

Jones, F. W. 1953. *Trends of Life.* Arnold, London.

Kauffman, E. G. 1984. The fabric of Cretaceous marine extinctions. In: Berggren, W. A. and Van Couvering, J. A. (eds.), *Catastrophes and Earth History.* Princeton University Press, Princeton, pp. 151–246.

Kauffman, E. G. 1986. High-resolution event stratigraphy: Regional and global Cretaceous bio-events. In: Walliser, O. H. (ed.), *Global Bio-Events.* Springer, Berlin, pp. 279–335.

Kauffman, S. A. 1985. Self-organization, selective adaptation, and its limits: A new pattern of inference in evolution and development. In: Depew, D. J. and Weber, B. H. (eds.), *Evolution at a Crossroads.* MIT Press, Cambridge, Mass., pp. 169–208.

Kaźmierczak, J. and Degens, E. T. 1986. Calcium and the early Eukaryotes. *Mitt. Geol.-Paläont. Inst. Univ. Hamburg* 61, 1–20.

Kaźmierczak, J.; Ittekot, V. and Degens, E. T. 1985. Biocalcification through time: Environmental challenge and cellular response. *Paläont. Z.* 59, 15–33.

Kelley, P. H. 1983. Evolutionary patterns of eight Chesapeake Group mollusks: Evidence for the model of punctuated equilibria. *J. Paleont.* 57, 581–598.

Kempe, S. and Degens, E. T. 1985. An early soda ocean? *Chem. Geol.* 53, 95–108.

Kent, D. V. 1977. An estimate of the duration of the faunal change at the Cretaceous-Tertiary boundary. *Geology* 5, 769–771.

Kettlewell, H. B. D. 1973. *The Evolution of Melanism.* Oxford University Press, Oxford.

Kimura, M. 1968. Evolutionary rate at the molecular level. *Nature* 217, 624–626.

Kimura, M. 1983. *The Neutral Theory of Molecular Evolution.* Cambridge University Press, Cambridge.

King, J. L. and Jukes, T. H. 1969. Non-Darwinian evolution. *Science* 164, 788–789.

Kirkpatrick, M. 1981. Quantum evolution and punctuated equilibria in continuous genetic characters. *Am. Nat.* 119, 833–848.

Kitchell, J. A. and Carr, T. R. 1985. Nonequilibrium model of diversification: Faunal turnover dynamics. In: Valentine, J. W. (ed.), *Phanerozoic Diversity Patterns.* Princeton University Press, Princeton, pp. 277–310.

Kitchell, J. A. and Estabrook, G. 1986. Was there 26-Myr periodicity of extinctions? [Discussion] *Nature* 321, 534–535.

Kitchell, J. A. and Hoffman, A. 1989. Rates of species-level origination and extinction: Functions of age, diversity, and history. In: Stenseth, N. C. (ed.), *Coevolution in Ecosystems,* in press. Cambridge University Press, Cambridge.

Kitchell, J. A. and Peña, D. 1984. Periodicity of extinction in the geological past: Deterministic versus stochastic explanations. *Science* 226, 689–692.

Kitcher, P. 1988. *Species.* MIT Press, Cambridge, Mass. (in press).

Koch, P. L. 1986. Clinal geographic variation in mammals: A test and implications for evolutionary theory. *Paleobiology* 12, 269–281.

Koestler, A. 1967. *The Ghost in the Machine.* Hutchinson, London.

Kovač, L. 1985. Calcium and *Saccharomyces cerevisiae. Biochim. Biophys. Acta* 840, 317–323.

Kowalewski, W. O. 1876. Monographie der Gattung *Anthracotherium* Cuv. und Versuch einer naturlichen Klassifikation der fossilen Huftiere. *Palaeontographica* 22, 31–346.

Kuhn, T. S. 1962. *The Structure of Scientific Revolutions.* University of Chicago Press, Chicago.

Kurtén, B. 1964. The evolution of the polar bear, *Ursus maritimus. Acta Zool. Fenn.* 108, 1–26.

Lakatos, I. 1970. Falsification and the methodology of scientific research programs. In: Lakatos, I. and Musgrave, A. (eds.), *Criticism and Growth of Knowledge.* Cambridge University Press, Cambridge, pp. 91–196.

Lamarck, J. B. P. A. 1809. *Philosophie Zoologique.* Dentu, Paris.

Lande, R. 1976. Natural selection and random genetic drift in phenotypic evolution. *Evolution* 30, 314–334.

Lande, R. 1986. The dynamics of peak shifts and the pattern of morphological evolution. *Paleobiology* 12, 343–354.

Larson, A. 1983. Neontological inferences of evolutionary pattern and process in the salamander family Plethodontidae. *Evol. Biol.* 17, 119–218.

Lazarus, D. 1986. Tempo and mode of morphologic evolution near the origin of the radiolarian lineage *Pterocanium prismatium. Paleobiology* 12, 175–189.

Leigh, E. G. 1977. How does selection reconcile individual advantage with the good of the group? *Proc. Nat. Acad. Sci. USA* 74, 4542–4546.

Levine, G. 1986. Darwin and the evolution of fiction. *New York Times Book Review,* October 5, pp. 1 and 60–61.

Levinton, J. S. 1982. Estimating stasis: Can a null hypothesis be too null? *Paleobiology* 8, 307.

Levinton, J. S. 1983. Stasis in progress: The empirical basis of macroevolution. *Ann. Rev. Ecol. Syst.* 14, 103–137.

Lewin, R. 1980. Evolutionary theory under fire. *Science* 210, 883–887.

Lewontin, R. C. 1983. Gene, organism, and environment. In: Bendall, D. S. (ed.), *Evolution from Molecules to Men.* Cambridge University Press, Cambridge, pp. 273–286.

Linné, K. 1735. *Systema Naturae.* Haak, Leiden.

Łomnicki, A. 1987. *Population Ecology of Individuals.* Princeton University Press, Princeton.

Łomnicki, A. and Hoffman, A. 1987. Poziom działania doboru naturalnego: dobro gatunku, dobór gatunków i dobór grupowy. *Kosmos* 36, 433–456.

Lovelock, J. E. 1979. *Gaia: A New Look at Life on Earth.* Oxford University Press, Oxford.

Løvtrup, S. 1982. The four theories of evolution. *Riv. Biol.* 75, 53–66, 231–272, 385–409.

Løvtrup, S. 1987. *Darwinism: The Refutation of a Myth.* Croom Helm, Beckenham.

Luria, S. E.; Gould, S. J. and Singer, S. 1981. *A View of Life.* Benjamin/Cummings, Menlo Park.

Lyell, C. 1832. *Principles of Geology.* Murray, London.

Lysenko, T. D. 1956. *O biologicheskom vide i videobrazovanii.* Nauka, Moskva.

MacArthur, R. H. and Wilson, E. O. 1967. *The Theory of Island Biogeography.* Princeton University Press, Princeton.

Maderson, P. F. A.; Alberch, P.; Goodwin, B. C.; Gould, S. J.; Hoffman, A.; Murray, J. D.; de Ricqlès, A.; Seilacher, A.; Wagner, G. P. and Wake, D. B. 1982. The role of development in macroevolutionary change. In: Bonner, J. T. (ed.): *Evolution and Development.* Springer, Berlin, pp. 279–312.

Malmgren, B. A.; Berggren, W. A. and Lohmann, G. P. 1983. Evidence for punctuated gradualism in the Late Neogene *Globorotalia tumida* lineage of planktonic foraminifera. *Paleobiology* 9, 377–389.

Marshall, H. T. 1928. Ultra-violet and extinction. *Amer. Natur.* 62, 165–187.

Martinell, J. and Hoffman, A. 1983. Species duration patterns in the Pliocene gastropod fauna of Emporda (Northeast Spain). *N. Jb. Geol. Palaont., Mh.* 1983, 698–704.

Maynard Smith, J. 1958. *The Theory of Evolution.* Penguin, Harmondsworth.

Maynard Smith, J. 1976. Group selection. *Quart. Rev. Biol.* 51, 277–283.

Maynard Smith, J. 1978. *The Evolution of Sex.* Cambridge University Press, Cambridge.

Maynard Smith, J. 1983. Current controversies in evolutionary biology. In: Grene, M. (ed.), *Dimensions of Darwinism.* Cambridge University Press, Cambridge, pp. 273–286.

Maynard Smith, J. 1986. *The Problems of Biology.* Oxford University Press, Oxford.

Maynard Smith, J.; Burian, R.; Kauffman, S.; Alberch, P.; Campbell, J.; Goodwin, B.; Lande, R.; Raup, D. and Wolpert, L. 1985. Developmental constraints and evolution. *Quart. Rev. Biol.* 60, 265–287.

Mayr, E. 1942. *Systematics and the Origin of Species.* Columbia University Press, New York.

Mayr, E. 1963. *Animal Species and Evolution.* Harvard University Press, Cambridge, Mass.

Mayr, E. 1982. Processes of speciation in animals. In: Barigozzi, C. (ed.), *Mechanisms of Speciation.* Liss, New York, pp. 1–20.

Mayr, E. 1986. Natural selection: The philosopher and the biologist. *Paleobiology* 12, 233–239.

Mayr, E. 1987. The ontological status of species: Scientific progress and philosophical terminology. *Biol. Philos.* 2, 145–166.

McKinney, M. L. 1985. Distinguishing patterns of evolution from patterns of deposition. *J. Paleont.* 59, 561–567.

McKinney, M. L. 1987. Taxonomic selectivity and continuous variation in mass and background extinctions of marine taxa. *Nature* 325, 143–145.

McLaren, D. J. 1970. Time, life, and boundaries. *J. Paleont.* 44, 301–315.

McLean, D. M. 1978. A terminal Mesozoic "Greenhouse": Lessons from the past. *Science* 201, 401–406.

McNamara, K. J. 1982. Heterochrony and phylogenetic trends. *Paleobiology* 8, 130–142.

Meyer-Abich, A. 1963. *Geistgeschichtliche Grundlagen der Biologie.* Fischer, Stuttgart.

Minkoff, E. C. 1983. *Evolutionary Biology.* Addison-Wesley, Reading, Mass.

Moore, R. C.; Lalicker, C. G. and Fischer, A. G. 1952. *Invertebrate Fossils.* McGraw-Hill, New York.

Morgan, T. H. 1903. *Evolution and Adaptation.* Macmillan, New York.

Mori, S. 1986. Changes of characters of *Drosophila melanogaster* brought about during life in constant darkness and considerations on the processes through which these changes were induced. *Zool. Science* 3, 945–957.

Muller, H. J. 1949. Reintegration of the symposium on genetics, paleontology, and evolution. In: Jeppson, G. L.; Mayr, E. and Simpson, G. G. (eds.), *Genetics, Paleontology, and Evolution.* Princeton University Press, Princeton, pp. 421–427.

Narkiewicz, N. and Hoffman, A. MS. Glacioeustasy at the Frasnian/Famennian transition?

Nelson, G. and Platnick, N. 1981. *Systematics and Biogeography.* Columbia University Press, New York.

Nelson, G. and Platnick, N. 1984. Systematics and evolution. In: Ho, M.-W. and Saunders, P. T. (eds.), *Beyond Neo-Darwinism.* Wiley, Chichester, pp. 143–158.

Newell, N. D. 1967. Revolutions in the history of life. *Geol. Soc. Amer. Spec. Paper* 89, 63–91.

Newman, C. M.; Cohen, J. E. and Kipnis, C. 1985. Neo-Darwinian evolution implies punctuated equilibria. *Nature* 315, 400–401.

Niklas, K. J.; Tiffney, B. H. and Knoll, A. H. 1980. Apparent changes in the diversity of fossil plants. *Evol. Biol.* 12, 1–89.

Niklas, K. J.; Tiffney, B. H. and Knoll, A. H. 1985. Patterns in vascular land plant diversification: An analysis at the species level. In: Valentine, J. W. (ed.), *Phanerozoic Diversity Patterns.* Princeton University Press, Princeton, pp. 97–128.

Nitecki, M. H. and Hoffman, A. (eds.) 1987. *Neutral Models in Biology.* Oxford University Press, New York.

Noma, E. and Glass, A. L. 1987. Mass extinction pattern: Result of chance. *Geol. Mag.* 124, 319–322.

O'Brien, S. J.; Nash, W. G.; Wildt, D. E.; Bush, M. E. and Benveniste, R. E. 1985. A molecular solution to the riddle of the giant panda's phylogeny. *Nature* 317, 140–144.

Odin, G. S. (ed.) 1982. *Numerical Dating in Stratigraphy*. Wiley, New York.

Officer, C. B.; Hallam, A.; Drake, C. L. and Devine, J. D. 1987. Late Cretaceous and paroxysmal Cretaceous/Tertiary extinctions. *Nature* 326, 143–149.

Oppel, A. 1856–1858. *Die Juraformation Englands, Frankreichs und des südwestlichen Deutschlands*. Elmer und Seubert, Stuttgart.

Orbigny, A. d'. 1842–1851. *Paléontologie Française; description des Mollusques et Rayonnés fossiles*. Mason, Paris.

Osborn, H. F. 1934. Aristogenesis, the creative principle in the origin of species. *Amer. Natur.* 68, 193–235.

Padian, K. and Clemens, W. A. 1985. Terrestrial vertebrate diversity: Episodes and insights. In: Valentine, J. W. (ed.), *Phanerozoic Diversity Patterns*. Princeton University Press, Princeton, pp. 41–96.

Palmer, A. R. 1983. Decade of North American Geology. 1983 Geologic Time Scale. *Geology* 11, 503–504.

Patterson, C. 1981. Significance of fossils in determining evolutionary relationships. *Ann. Rev. Ecol. Syst.* 12, 195–223.

Patterson, C. 1982. Morphological characters and homology. In: Joysey, K. A. and Friday, A. E. (eds.), *Problems of Phylogenetic Reconstruction*. Academic Press, London, pp. 21–74.

Patterson, C. and Smith, A. B. 1987. Is the periodicity of extinctions a taxonomic artefact? *Nature* 330, 248–251.

Paul, C. R. C. 1982. The adequacy of the fossil record. In: Joysey, K. A. and Friday, A. E. (eds.), *Problems of Phylogenetic Reconstruction*. Academic Press, London, pp. 75–117.

Pavlova, M. V. 1924. *Prichiny Vymiranya Zhivotnykh v Proshedshye Geologicheskie Periody*. Nauka, Moskva.

Penny, D. 1983. Charles Darwin, gradualism, and punctuated equilibrium. *Syst. Zool.* 32, 72–74.

Penny, D.; Foulds, L. R. and Hendy, M. D. 1982. Testing the theory of evolution by comparing phylogenetic trees constructed from five different protein sequences. *Nature* 297, 197–200.

Petry, D. 1982. The pattern of phyletic speciation. *Paleobiology* 8, 56–66.

Philiptschenko, J. 1927. *Variabilität und Variation*. Borntraeger, Berlin.

Phillips, J. 1860. *Life on the Earth*. Macmillan, London.

Pilbeam, D. R. 1972. *The Ascent of Man*. Macmillan, London.

Pollard, J. W. (ed.) 1984. *Evolutionary Theory: Paths into the Future*. Wiley, Chichester.

Popper, K. R. 1959. *The Logic of Scientific Discovery*. Hutchinson, London.

Popper, K. R. 1963. *Conjectures and Refutations: The Growth of Scientific Knowledge*. Harper & Row, New York.

Prothero, D. R. 1985. North American mammalian diversity and Eocene-Oligocene extinctions. *Paleobiology* 11, 389–405.

Prothero, D. R. and Lazarus, D. 1980. Planktonic microfossils and the recognition of ancestors. *Syst. Zool.* 29, 119–129.

Quenstedt, F. A. 1852. *Handbuch der Petrefaktenkunde.* Laupp'sche Buchhandlung, Tübingen.

Quine, W. V. O. 1980. *From a Logical Point of View.* Harvard University Press, Cambridge, Mass.

Raup, D. M. 1972. Taxonomic diversity during the Phanerozoic. *Science* 177, 1065–1071.

Raup, D. M. 1978. Cohort analysis of generic survivorship. *Paleobiology* 4, 1–15.

Raup, D. M. 1979. Size of the Permo-Triassic bottleneck and its evolutionary implications. *Science* 206, 217–218.

Raup, D. M. 1983. On the early origins of major biologic groups. *Paleobiology* 9, 107–115.

Raup, D. M. 1986. *The Nemesis Affair.* Norton, New York.

Raup, D. M. and Crick, R. E. 1981. Evolution of single characters in the Jurassic ammonite *Kosmoceras. Paleobiology* 7, 200–215.

Raup, D. M. and Sepkoski, J. J. 1984. Periodicity of extinctions in the geologic past. *Proc. Nat. Acad. Sci. USA* 81, 801–805.

Raup, D. M. and Sepkoski, J. J. 1986. Periodic extinction of families and genera. *Science* 231, 833–836.

Raup, D. M. and Stanley, S. M. 1978. *Principles of Paleontology,* 2nd ed. Freeman, San Francisco.

Reid, R. G. B. 1985. *Evolutionary Theory: The Unfinished Synthesis.* Croom Helm, Beckenham.

Reinecke, I. C. M. 1818. *Maris Protogaei Nautilos et Argonautas.* Ahl, Coburg.

Rensch, B. 1947. *Neuere Probleme der Abstammungslehre.* Ferdinand Enke, Stuttgart.

Rhodes, F. H. T. 1983. Gradualism, punctuated equilibrium and *The Origin of Species. Nature* 305, 269–272.

Ricoeur, P. 1979. *Histoire et Verité,* 3rd ed. Seuil, Paris.

Rieppel, O. 1986. Species are individuals: A review and critique of the argument. *Evol. Biol.* 20, 283–317.

Rieppel, O. 1987. Punctuational thinking at odds with Leibniz—and Darwin. *N. Jb. Geol. Paläont., Abh.* 174, 123–133.

Rosen, R. (ed.) 1985. *Theoretical Biology and Complexity.* Academic Press, Orlando, Fla.

Rosenberg, A. 1985. *The Structure of Biological Science.* Cambridge University Press, Cambridge.

Rosenzweig, M. L. 1975. On continental steady states of species diversity. In: Cody, M. L. and Diamond, J. M. (eds.), *Ecology and Evolution of Communities.* Belknap Press, Cambridge, Mass., pp. 121–140.

Ross, S. 1987. Are mass extinctions really periodic? *Probab. Engineer. Inform. Sci.* 1, 61–64.

Ruse, M. 1973. *The Philosophy of Biology.* Hutchinson, London.

Sadler, P. M. 1981. Sediment accumulation rates and the completeness of stratigraphic sections. *J. Geol.* 89, 569–584.

Salthe, S. N. 1985. *Evolving Hierarchical Systems.* Columbia University Press, New York.

Sarich, V. and Wilson, A. C. 1967. Immunological time scale for hominid evolution. *Science* 158, 1200–1203.

Schindewolf, O. H. 1950. *Grundfragen der Palaeontologie.* Schweizerbart'sche, Stuttgart.

Schindewolf, O. 1954a. Über die Faunenwende vom Paläozoikum zum Mesozoikum. *Z. Dtsch. Geol. Ges.* 105, 153–182.

Schindewolf, O. 1954b. Über die möglichen Ursachen der grossen urgeschichtlichen Faunenschnitte. *N. Jb. Geol. Paläont., Mh.* 1954, 457–465.

Schmalhausen, I. I. 1949. *The Factors of Evolution.* Blakiston, Philadelphia.

Schopf, T. J. M. 1974. Permo-Triassic extinctions: Relation to seafloor spreading. *J. Geol.* 82, 129–143.

Schopf, T. J. M. 1979. Evolving paleontological views on deterministic and stochastic approaches. *Paleobiology* 5, 337–352.

Schopf, T. J. M. 1981. Evidence from findings of molecular biology with regard to the rapidity of genomic change: Implications for species durations. In: Niklas, K. J. (ed.), *Paleobotany, Paleoecology and Evolution.* Praeger, New York, vol. 1, pp. 135–192.

Schopf, T. J. M. 1984. Climate is only half the story in the evolution of organisms through time. In: Brenchley, P. J. (ed.), *Fossils and Climate,* Wiley, Chichester, pp. 279–290.

Schweyen, R. J.; Wolf, K. and Kaudewitz, F. (eds.) 1983. *Mitochondria 1983: Nucleo-Mitochondrial Interactions.* De Gruyter, Berlin.

Sepkoski, J. J. 1978. A kinetic model of Phanerozoic taxonomic diversity. I. Analysis of marine orders. *Paleobiology* 4, 223–251.

Sepkoski, J. J. 1979. A kinetic model of Phanerozoic taxonomic diversity. II. Early Phanerozoic families and multiple equilibria. *Paleobiology* 5, 222–251.

Sepkoski, J. J. 1981. A factor analytic description of the Phanerozoic marine fossil record. *Paleobiology* 7, 36–53.

Sepkoski, J. J. 1982. A compendium of fossil marine families. *Milwaukee Publ. Mus. Contrib. Biol. Geol.* 51, 1–125.

Sepkoski, J. J. 1984. A kinetic model of Phanerozoic taxonomic diversity. III. Post-Paleozoic families and mass extinctions. *Paleobiology* 10, 246–267.

Sepkoski, J. J. 1986. Phanerozoic overview of mass extinctions. In: Raup,

D. M. and Jablonski, D. (eds.), *Patterns and Processes in the History of Life*. Springer, Berlin, pp. 277–296.

Sepkoski, J. J.; Bambach, R. K.; Raup, D. M. and Valentine, J. W. 1981. Phanerozoic marine diversity and the fossil record. *Nature* 293, 435–437.

Sepkoski, J. J. and Hulver, M. L. 1985. An atlas of Phanerozoic clade diversity diagrams. In: Valentine, J. W. (ed.), *Phanerozoic Diversity Patterns*. Princeton University Press, Princeton, pp. 11–39.

Sepkoski, J. J. and Raup, D. M. 1986. Periodicity in marine extinction events. In: Elliott, D. K. (ed.), *Dynamics of Extinction*. Wiley, New York, pp. 3–36.

Sheldon, P. R. 1987. Parallel gradualistic evolution of Ordovician trilobites. *Nature* 330, 561–563.

Sheng, J. Z.; Chen, C. Z.; Wang, Y. G.; Rui, L.; Liao, Z. T.; Bando, Y.; Ishi, K.; Nakazawa, K. and Nakamura, K. 1984. Permian-Triassic boundary in middle and eastern Tethys. *J. Fac. Sci. Hokkaido Univ. IV* 21, 133–181.

Signor, P. W. 1985. Real and apparent trends in species richness through time. In: Valentine, J. W. (ed.), *Phanerozoic Diversity Patterns,* Princeton University Press, Princeton, pp. 129–150.

Simon, W. 1958. Erdgeschehen und Stammesgeschichte. *Geologie* 7, 808–825.

Simpson, G. G. 1944. *Tempo and Mode in Evolution*. Columbia University Press, New York.

Simpson, G. G. 1953. *The Major Features of Evolution*. Columbia University Press, New York.

Simpson, G. G. 1960. The history of life. In: Tax, S. (ed.), *Evolution After Darwin*. University of Chicago Press, Chicago, pp. 117–180.

Slobodkin, L. B. 1987. How to be objective in community studies. In: Nitecki, M. H. and Hoffman, A. (eds.), *Neutral Models in Biology*. Oxford University Press, New York, pp. 93–108.

Snelling, N. J. (ed.) 1985. The chronology of the geologic record. *Geol. Soc. London Mem.* 10.

Sober, E. 1975. *Simplicity*. Oxford University Press, Oxford.

Sober, E. 1984. *The Nature of Selection*. MIT Press, Cambridge, Mass.

Sobolev, D. N. 1928. *Zemla i Zhizn. O Prichinakh Vymiranya Organizmov*. Kiev.

Sondaar, P. Y. 1977. Insularity and its effect on mammlian evolution. In: Hecht, M. K.; Goody, P. C. and Hecht, B. M. (eds.), *Major Patterns in Vertebrate Evolution*. Plenum Press, New York, pp. 671–708.

Sorhannus, U.; Fenster, E. J.; Burckle, L. H. and Hoffman, A. 1988. Cladogenetic and anagenetic changes in the morphology of *Rhizosolenia praebergonii* Mukhina. *Hist. Biol.* 1, in press.

Stanley, S. M. 1975. A theory of evolution above the species level. *Proc. Nat. Acad. Sci. USA* 72, 646–650.

Stanley, S. M. 1979. *Macroevolution—Pattern and Process*. Freeman, San Francisco.

Stanley, S. M. 1982a. Speciation and the fossil record. In: Barigozzi, C. (ed.), *Mechanisms of Speciation*. Liss, New York, pp. 41–50.

Stanley, S. M. 1982b. Macroevolution and the fossil record. *Evolution* 36, 460–473.

Stanley, S. M. 1982c. Gastropod torsion: Predation and the opercular imperative. *N. Jb. Geol. Palaont., Abh.* 164, 95–107.

Stanley, S. M. 1986. Population size, extinction, and speciation: The fission effect in Neogene Bivalvia. *Paleobiology* 12, 89–110.

Stanley, S. M. 1987. *Extinction*. Freeman, New York.

Stanley, S. M. and Newman, W. A. 1980. Competitive exclusion in evolutionary time: The case of acorn barnacles. *Paleobiology* 6, 173–183.

Stanley, S. M.; Van Valkenburgh, B. and Steneck, R. S. 1983. Coevolution and the fossil record. In: Futuyma, D. J. and Slatkin, M. (eds.), *Coevolution*. Sinauer, Sunderland, Mass., 328–349.

Stanley, S. M. and Yang, X. 1987. Approximate evolutionary stasis for bivalve morphology over millions of years: A multivariate, multi-lineage study. *Paleobiology* 13, 113–139.

Stearns, S. C. 1982. The role of development in the evolution of life-histories. In: Bonner, J. T. (ed.), *Evolution and Development*. Springer, Berlin, pp. 237–258.

Stebbins, G. L. 1950. *Variation and Evolution in Plants*. Columbia University Press, New York.

Stebbins, G. L. 1982. Perspectives in evolutionary theory. *Evolution* 36, 1109–1118.

Stebbins, G. L. 1987. Species concepts: Semantics and actual situations. *Biol. Philos.* 2, 198–203.

Steele, E. J. 1981. *Somatic Selection and Adaptive Evolution*. University of Chicago Press, Chicago.

Stenseth, N. C. and Maynard Smith, J. 1984. Coevolution in ecosystems: Red Queen evolution or stasis? *Evolution* 38, 870–880.

Stepanov, D. L. 1959. Neokatastrofizm v paleontolgii nashikh dney. *Paleont. Zh.* 1959 (4), 11–16.

Stigler, S. M. 1987. Testing hypotheses or fitting models? Another look at mass extinctions. In: Nitecki, M. H. and Hoffman, A. (eds.), *Neutral Models in Biology*. Oxford University Press, New York, pp. 147–159.

Stigler, S. M. and Wagner, M. W. 1987. A substantial bias in nonparametric tests for periodicity in geophysical data. *Science* 238, 940–945.

Strong, D. R. 1985. Density vagueness: Abiding the variance in the demography of real populations. In: Diamond, J. and Case, T. (eds.), *Community Ecology*. Harper & Row, New York, pp. 257–268.

Strong, D. R. and Rey, J. R. 1982. Testing for MacArthur-Wilson equilibrium with the arthropods of the miniature *Spartine* archipelago at Oyster Bay, Florida. *Amer. Zool.* 22, 355–360.

Strong, D. R.; Simberloff, D.; Abele, L. G. and Thistle, A. B. (eds.) 1984. *Ecological Communities: Conceptual Issues and the Evidence.* Princeton University Press, Princeton.

Szarski, H. 1986. *Mechanizmy ewolucji,* 3rd ed. Państowe Wydawnictwo Naukowe, Warszawa.

Teilhard de Chardin, P. 1955. *Le Phenomène Humain.* Seuil, Paris.

Thaler, L. 1983. Image paléontologique et contenu biologique des lignées évolutives. In: Chaline, J. (ed.), *Modalités, Rhythmes et Mécanismes de l'Evolution Biologique.* CNRS, Paris, pp. 327–336.

Thomson, K. S. 1982. The meanings of evolution. *Amer. Sci.* 70, 529–531.

Towe, K. M. 1986. Fossil Preservation. In: Boardman, R. S.; Cheetham, A. H. and Rowell, A. J. (eds.), *Fossil Invertebrates.* Freeman, San Francisco, pp. 36–41.

Turner, J. R. G. 1981. Adaptation and evolution in *Heliconius:* A defense of neo-Darwinism. *Ann. Rev. Ecol. Syst.* 12, 99–121.

Turner, J. R. G. 1986. The genetics of adaptive radiation: A neo-Darwinian theory of punctuational evolution. In: Raup, D. M. and Jablonski, D. (eds.), *Patterns and Processes in the History of Life.* Springer, Berlin, pp. 183–207;

Urbanek, A. 1963. On generation and regeneration of cladia in some Upper Silurian monograptids. *Acta Palaeont. Polon.* 8, 135–258.

Valentine, J. W. 1970. How many marine invertebrate fossil species? A new approximation. *J. Paleont.* 44, 410–415.

Valentine, J. W. 1980. Determinants of diversity in higher taxonomic categories. *Paleobiology* 6, 444–450.

Valentine, J. W.; Foin, T. C.; and Peart, D. 1978. A provincial model of Phanerozoic marine diversity. *Paleobiology* 4, 55–66.

Vandel, A. 1964. *Biospéologie.* Gallimard, Paris.

Van Valen, L. 1973. A new evolutionary law. *Evol. Theory* 1, 1–30.

Van Valen, L. 1975. Group selection, sex, and fossils. *Evolution* 29, 87–94.

Van Valen, L. 1976. Individualistic classes. *Philos. Sci.* 43, 539–541.

Van Valen, L. M. 1982. Why misunderstand the evolutionary half of biology? In: Saarinen, E. (ed.), *Conceptual Issues in Ecology.* Reidel, Dordrecht, pp. 323–344.

Van Valen, L. M. 1985. How constant is extinction? *Evol. Theory* 7, 93–106.

Van Valen, L. M. and Maiorana, V. C. 1985. Patterns of origination. *Evol. Theory* 7, 107–125.

Vermeij, G. J. 1977. The Mesozoic marine revolution: evidence from snails, predators, and grazers. *Paleobiology* 3, 245–258.

Vermeij, G. J. 1987. *Evolution as Escalation.* Princeton University Press, Princeton.

Vernadsky, V. A. 1930. *Geochemie in ausgewählten Kapiteln.* Akademische Verlag, Leipzig.

Vrba, E. S. 1980. Evolution, species, and fossils: How does life evolve? *S. Afr. J. Sci.* 76, 61–84.

Vrba, E. S. 1982. Darwinism in 1982: The triumph and the challenges. *S. Afr. J. Sci.* 76, 61–84.

Vrba, E. S. 1983. Macroevolutionary trends: New perspectives on the roles of adaptation and incidental effect. *Science* 221, 387–389.

Vrba, E. S. 1985a. Environment and evolution: Alternative causes of the temporal distribution of evolutionary events. *S. Afr. J. Sci.* 81, 229–236.

Vrba, E. S. 1985b. Introductory comments on species and speciation. In: Vrba, E. S. (ed.), *Species and Speciation.* Transvaal Museum, Pretoria, pp. ix–xviii.

Vrba, E. S. and Eldredge, N. 1984. Individuals, hierarchies, and processes: Towards a more complete evolutionary theory. *Paleobiology* 10, 146–171.

Vrba, E. S. and Gould, S. J. 1986. The hierarchical expansion of sorting and selection: Sorting and selection cannot be equated. *Paleobiology* 12, 217–226.

Waagen, E. 1869. Die Formenreihe des Ammonites subradiatus. Versuch einer paläontologischen Monographie. *Geognostisch-Paläontologische Beiträge* 2 (2), 181–256.

Waddington, C. H. 1957. *The Strategy of the Genes.* Allen & Unwin, London.

Wake, D. B.; Roth, G.; Wake, M. H. 1983. On the problem of stasis in organismal evolution. *J. Theor. Biol.* 101, 211–224.

Webster, G. C. and Goodwin, B. C. 1982. The origin of species: A structuralist approach. *J. Social Biol. Struct.* 5, 15–47.

White, M. 1965. *Foundations of Historical Knowledge.* Harper & Row, New York.

Whittaker, R. H. 1977. Evolution of species diversity in land communities. *Evol. Biol.* 10, 1–67.

Wilkins, H. 1971. Genetic interpretation of regressive evolutionary processes: Studies on hybrid eyes of two *Astyanax* (Characidae, Pisces). *Evolution* 25, 530–544.

Williams, G. C. 1966. *Adaptation and Natural Selection.* Princeton University Press, Princeton.

Williams, N. E. 1984. An apparent disjunction between the evolution of form and substance in the genus *Tetrahymena. Evolution* 38, 25–33.

Williamson, P. G. 1981. Paleontological documentation of speciation in Cenozoic mollusks from Turkana Basin. *Nature* 293, 437–443.

Williamson, P. G. 1985. Punctuated equilibrium, morphological stasis and the paleontological documentation of speciation: A reply to Fryer,

Greenwood and Peake's critique of the Turkana Basin mollusk sequence. *Biol. J. Linn. Soc.* 26, 307–324.

Williamson, P. G. 1987. Selection or constraint?: A proposal on the mechanism of stasis. In: Campbell, K. S. W. and Day, M. F. (eds.), *Rates of Evolution*. Allen & Unwin, London, pp. 129–142.

Wilson, D. S. 1980. *The Natural Selection of Populations and Communities*. Benjamin/Cummings, Menlo Park.

Wilson, E. O. 1975. *Sociobiology: The New Synthesis*. Harvard University Press, Cambridge, Mass.

Wilson, E. O. 1978. *On Human Nature*. Harvard University Press, Cambridge, Mass.

Wilson, E. O. 1984. *Biophilia*. Harvard University Press, Cambridge, Mass.

Winston, J. E. and Cheetham, A. H. 1984. The bryozoan *Nellia tenella* as a living fossil. In: Eldredge, N. and Stanley, S. M. (eds.), *Living Fossils*. Springer, New York, pp. 257–265.

Wintrebert, P. 1962. *Le Vivant, Créateur de son Evolution*. Masson, Paris.

Wolpoff, M. H. 1983. *Ramapithecus* and human origins: An anthropologist's perspective of changing interpretations. In: Ciochon, R. L. and Corruccini, R. S. (eds.), *New Interpretations of Ape and Human Ancestry*. Plenum Press, New York, pp. 651–676.

Woods, H. 1893. *Elementary Paleontology*. Cambridge University Press, Cambridge.

Wright, S. 1931. Evolution in Mendelian populations. *Genetics* 16, 97–159.

Wright, S. 1945. Tempo and mode in evolution: A critical review. *Ecology* 26, 415–419.

Wright, S. 1970–1978. *Evolution and the Genetics of Populations*. University of Chicago Press, Chicago.

Wright, S. 1982. The shifting balance theory and macroevolution. *Ann. Rev. Gen.* 16, 1–19.

Wynne-Edwards, V. C. 1962. *Animal Dispersion in Relation to Social Behavior*. Oliver & Boyd, Edinburgh.

Yin, H. F. 1985. Bivalves near the Permian-Triassic boundary in South China. *J. Paleont.* 59, 572–600.

Zinsmeister, W. J. and Feldmann, R. M. 1984. Cenozoic high latitude heterochroneity of Southern Hemisphere marine faunas. *Science* 224, 281–283.

Zittel, K. von. 1876–1893. *Handbuch der Paläontologie*. Oldenbourg, München.

Index